IEE Software Engineering Series 1
Series Editor: D.A.H. Jacobs

Integrated project support environments

Integrated project support environments

Edited by
John McDermid

Peter Peregrinus Ltd on behalf of the Institution of Electrical Engineers

Published by: Peter Peregrinus Ltd., London, UK.

© 1985 Peter Peregrinus Ltd.

ISBN 0 86341 050 2

Printed in England by Short Run Press Ltd., Exeter

Contents

Foreword

This volume contains the proceedings of the conference on Integrated Project Support Environments (IPSEs) held at the University of York from April 10th to April 12th 1985.

The purpose of the conference was to discuss the current state of research and development on IPSEs, and to provide a forum for interchange of ideas for people working in the field. The conference was attended by over 100 people and papers were presented by workers from France, the UK and the USA.

The papers cover a variety of topics, but the bulk of them describe IPSEs which are under development or which are in use in research laboratories.

The conference would not have been possible without the efforts of a great number of people. I wish to thank the authors for preparing and presenting their papers. I am grateful to the other members of the organising committee: Mike Falla, Howard Nicholls, Brian Randell and Ian Wand for their help in selecting papers and organising the conference. Thanks are due to the Alvey Directorate for financial support.

Finally, I wish particularly to thank Ian Wand and Jenny Turner who were responsible for the local arrangements and for making the conference a social, as well as a technical, success.

John McDermid.
Hook, May 1985.

Introduction

Software development comprises a number of activities including requirements analysis, system design, implementation and testing. At present few software development projects have the benefit of systematic, computer based, support for all these activities. Indeed most projects employ only a small number of loosely related tools concentrated around the programming activity.

The information manipulated by the tools is usually stored as files on some host computer. It is rare for information to be kept concerning the relationships between the files so, for example, it is rare to be able to trace part of an implementation back to the requirement which it satisfies. This means that requirements and design specifications (where they exist) frequently do not accurately reflect changes in the implementation. This leads to problems in software maintenance.

Further, in large scale software developments these files tend to be shared between different members of the project team, different system releases, and so on. This leads to problems of version and variant management, change control and configuration control. These problems are usually addressed by ad hoc file naming conventions and by the use of particular file directory structures.

Integrated Project Support Environments (IPSEs) are intended to address the problems outlined above by providing a set of tools covering the whole software life cycle, and supporting many different roles in the development process, within a coherent framework. The framework, or infrastructure, holds the information necessary to allow the problems of configuration control, change control, and so on to be solved more systematically. In other words an IPSE provides the working environment for the professional Software Engineer.

The term IPSE was introduced fairly recently, but work on environments has been going on for some time. CADES[1], developed by ICL in the mid 1970s to support operating system development, was an early environment with many of the characteristics of an IPSE. In 1980, a document, known as STONEMAN[2], was published setting out the requirements for an APSE (an IPSE for Ada). A considerable amount of fundamental work on IPSEs, of both a theoretical and practical nature, has been carried out as part of the Ada programme, e.g. [3]. This experience is now being applied in the development of environments for other languages.

At present there are a small number of IPSE products commercially available, and there is considerable research and development activity in Universities and in the software industry. It was this wide range of activity, and the desire to foster communication amongst those working on IPSEs, which prompted the organisation of the conference. The organisers believe that this wide range of activity is reflected in the papers presented at the conference.

DSEE, which is described in chaper 2, is a commercially available product running on the Apollo Domain. The UNIX tools described in chapter 5 were produced for use in UK Universities.

A number of environments are being used in research laboratories, e.g. Cedar (chapter 1), Flex (chapter 6), SMALLTALK (chapter 10) and MENTOR (chapter 14). All of these environments are now being used outside their original laboratories, but they are not yet in widespread use.

Most of the papers describe major research and development projects aimed at producing prototype IPSEs or commercial products. These projects have rather different emphasis. The work on MULE (chapter 3) and SPRAC (chapter 9) is aimed at supporting formal specifications. The environment described by Sellars (chapter 13) is intended for use in commercial data processing. Both ASPECT (chapter 7) and ECLIPSE (chapter 8) are intended to support geographically distributed working and to be open to the introduction of new tools, and the use of existing tools.

    Research topics include the use of abstract data types
(chapter 11) and knowledge based approaches (chapter 4) in
the IPSE infrastructure. The use of abstract data types for
fast prototyping is discussed in chapter 12.

    The final paper summarises the findings of an
Alvey-funded project to assess the requirements for future
generations of IPSEs.

John McDermid.
Hook, May 1984.

[1]   "CADES - software engineering in practice",
        R.W. McGuffin, A.E. Elliston, B.R. Tranter,
        P.N. Westmacott, Proc. 4th Int. Cont. on
        Software Engineering, 1979.

[2]   "Stoneman : Requirements for Ada Programming
        Support Environments", Department of
        Defence, 1980.

[3]   "Life Cycle Support in the Ada Environment",
        J.A. McDermid, K. Ripken,
        Cambridge University Press, 1984.

List of contributors

Chapter 1
James Donahue
Xerox Palo Alto Research Center, USA

Chapter 2
D.B. Leblang and G. McLean
Apollo Computers Inc.
Chelmsford
Ma 01824, USA

Chapter 3
I.D. Cottam, C.B. Jones, T. Nipkow, A.C. Wills,
M.I. Wolczko and A. Yaghi
Dept. of Computer Science
Manchester University, UK

Chapter 4
O.P. Brereton and P. Singleton
Dept. of Computer Science
University of Keele, UK

Chapter 5
D.J. Barnes, J.D. Bovey, P.J. Brown, H.P. Siemon
Computing Laboratory
University of Kent at Canterbury, UK

Chapter 6
I.F. Currie
RSRE
Great Malvern, UK

Chapter 7
J.A. Hall
Systems Designers
Camberley, UK
P. Hitchcock
Newcastle University, UK
R. Took
University of York, UK

Chapter 8
A. Alderson, M.E. Falla
Software Sciences Ltd.
Macclesfield, UK
M.F. Bott
University of Wales
Aberystwyth, UK

Chapter 9
J. Foisseau, R. Jacquart, M. Lemaitre,
M. Lemoine and G. Zanon
Dept. d'études et de recherches en informatique
de Toulouse
Toulouse
FRANCE

Chapter 10
L.P. Deutsch
Xerox Palo Alto Research Center
USA

Chapter 11
A.R. Jackson
Dept. of Computer Studies & Mathematics
Huddersfield Polytechnic
UK

Chapter 12
J.W. Hughes and M.S. Powell
UMIST
Manchester
UK

Chapter 13
P.W. Sellars
BIS Applied Systems Ltd.
London
UK

Chapter 14
B. Lang
INRIA
Domaine de Voluceau-Rocquencourt
Le Chesnay
FRANCE

Chapter 15
R.A. Snowdon, N.C. Munro
ICL
Stoke-on-Trent, UK
N.W. Davis
STC/IDEC
Harlow, UK
M.I. Jackson
STL
Harlow, UK

Chapter 1

Cedar:
An environment for 'experimental' programming

James Donahue

ABSTRACT

Cedar, an environment for "experimental programming," has been under development in the Computer Science Laboratory at the Xerox Palo Alto Research Center for the past several years. Aspects of the environment are discussed in Deutsch (4), Teitelman (30 and 31), Brown (3), Donahue (6), Swinehart (25), Rovner (20). This paper presents a discussion of what "experimental programming" is and how the Cedar environment supports this activity.

1.1 INTRODUCTION

Cedar is an environment for "experimental programming" that has been under development in the Computer Science Laboratory at the Xerox Palo Alto Research Center for the past several years. Teitelman's CSL technical report (Teitelman (30)) gives a detailed description of Cedar from the programmer's perspective; additional aspects of the environment are discussed in Deutsch (4), Teitelman (31), Brown (3), Donahue (6), Swinehart (25), and Rovner (20).

The Cedar project was not intended to be a project to do research in programming environments; thus, many parts of it are relatively conservative in design. Instead, Cedar was intended to be a research tool that would make the system development done in the lab much easier to do. So to understand the design of Cedar it is first necessary to understand the prevailing view of systems research in CSL — what we wanted the tool for. Then we will discuss some of the characteristics of Cedar that make it useful for our purposes.

1.2 EXPERIMENTAL PROGRAMMING

One of the basic paradigms for computer research in CSL has been to construct prototype systems that are solid enough to be used by moderate numbers of people as part of their daily activities. Some of the results of the work done in CSL have proved to be major successes: for instance, the Alto computer (Thacker (32)) and the Grapevine system (Birrell (2)) were widely distributed throughout the Xerox Corporation and became part of the day-to-day working environment of a large number of people. While few successes are of the magnitude of Alto or Grapevine, building systems that get substantial use is an important part of the research activity in the lab. We feel that many interesting aspects of systems design do not manifest themselves until one gives systems the "stress test" of heavy use.

The production of systems of this sort is what we mean by *experimental programming*. To quote from the report outlining the proposed requirements for Cedar (Deutsch (4)), "experimental programming is the production of moderate-sized systems that are usable by moderate numbers of people in order to test ideas about such systems." One way to understand the requirements of experimental programming environments is to contrast this activity with other possible styles of program development.

One programming style common to research labs and universities is *exploratory programming*. (Sheil (27)). By this we mean the development of primarily one-person, one-of-a-kind software that is used by the developer to solidify some systems idea. The system being produced is generally not intended for wide distribution; in fact, it may never be completed beyond the point of determining the validity of the concept. The most important characteristic of a programming environment for doing exploratory programming is probably the speed at which a system can be put together. Thus, being able to delay decisions (perhaps putting them off indefinitely) and being able to make small changes quickly are critical. The various Lisp environments are generally described as environments for this type of programming.

At the other end of the spectrum, one finds what I will call *production programming*. Production programming is most easily characterized as the development, probably by a large group, of a software system that is intended to be a major capital asset for the developer. For instance, the online transaction software developed by a bank is critical to the business success of the organization and is as much a capital asset as the bank's buildings. Here the most important characteristic of an environment is not the speed with which a system can be constructed, but the coordination of large numbers of people and the ease of program maintenance. It is less important to do things quickly than it is to do things methodically. Thus, it is important to have version control systems, to record the change history of programs, to have test suites, etc.

Experimental programming falls between these two extremes. While speed of construction is important, it is no more important than the ease of long-term maintenance. Groups tend to be small, so some of the tools necessary to manage large groups can be dispensed with. However, the desire to have a moderately sized user community means that stability of the environment is an important matter; you can't always be pulling the rug out from under your users! This means that the ability to make *incremental* changes to a system is of critical concern.

Making incremental changes efficiently involves two costs. The first is clearly the cost of the change itself. Like environments for exploratory programming, we want to be able to make these changes quickly; once an implementor knows what needs to be changed, it should be possible to make the necessary changes and rebuild the system in a very short time. The second cost is ensuring that a proposed change is *safe*, that the scope of its effect is limited to the components that are being changed. Users need such assurances when adding new (or recently modified) software components to their working environments; implementors need these same assurances both to be able to attract new users (who demand stability) and because the implementors are themselves users of the system. In this regard, experimental programming has much in common with production programming.

Cedar was developed as an environment to facilitate experimental programming: this paper explores some of the successes (and failures) of Cedar in this regard. We will first discuss some of the features of the environment that facilitate quick changes. These aspects of Cedar are not particularly novel: workstations give the programmer guaranteed computational resources and windowing environments make it easy to move among several parallel tasks. Additionally, Cedar has an integrated document/program preparation that helps make programs easier to read and to write, so that finding the right things to change can be made much easier.

More important, however, are the tools for managing incremental change. The most important considerations in managing change are that the assumptions upon which a software system depends must be explicit in the code (since the code is the only repository of trustworthy information about a system) and that these assumptions be checked by all components of the underlying environment. If a new component violates assumptions made by other parts of a running system, then it must not be possible to run the new component. The second part of this paper is a description of the parts of the Cedar environment that assist in managing incremental change to the system.

The Cedar programming language is a superset of the Mesa language; thus, the same clear separation of interface and implementation that one finds in Mesa is present in Cedar. This is undoubtedly the most important means of documenting the structure of a system in the code. Additionally, the version control system, *DF files*, that is shared between the Mesa and Cedar environments makes it easy for programmers to store and retrieve consistent versions of the collection of files that make up a component.

The major extension of Cedar from Mesa is the inclusion of *collectible storage*, so that the programmer is no longer completely responsible for the management of free storage. Since Cedar also includes low-level constructs whose careless use can destroy the storage invariants, the Cedar language includes a *safe subset*, a subset of the language that can be recognized by the compiler and in which no operations can destroy the collector invariants. Thus, programs written in the safe subset are known not to be potential causes of storage smashes.

1.3 MAKING CHANGES QUICKLY

No matter what sort of programming one is doing, being able to make quick changes (once it is clear what needs to be changed) is facilitated by having a computational environment with fast (and guaranteed) response times and the ability to switch quickly among several tasks. Also, being able to use an integrated document and program composition system makes it easier to understand the code in the system. This makes it easier to determine how changes should be made.

(We mention these points not because they are novel in Cedar, but because they are major contributors to the productivity of programmer's using the environment. They are important considerations in the design and evaluation of any programming environment.)

1.3.1 Workstations

Certainly the most significant property of a programming environment is the amount of computing resources it puts at the user's disposal. One of the major advantages of the workstation model of computing is that a workstation provides the user with guaranteed computational resources (the machines don't run any faster at night). Cedar runs on the collection of Xerox workstations, the 1100 (or Dolphin), the 1108 (or Dandelion), and the 1132 (or Dorado) processors, with most of its users having Dandelions or Dorados. The Dandelion processor (also used in the Xerox Star system) runs at about 1/2 the speed of a VAX 11/780 for typical processing (Harslem (9)) and the Dorado is a very high-performance personal computer (it has a cycle time of 60ns (Pier (18))).

The original plans for the Cedar environment anticipated being able to make small program changes and rebuild a running system in a few seconds. The current system does not meet this rather ambitious goal (Teitelman (30)) states that a very modest change can be made to a Cedar program, the compiler invoked, and the new program run again in less than one minute). Nevertheless, the impression of people using Cedar is that its performance on a Dorado makes a qualitative difference in the way one thinks about the difficulties of performing a programming task. Being able to get things done quickly increases our ambition to build systems.

1.3.2 Doing Many Things Simultaneously

Having plenty of computing resources means that one can spend more time thinking and less time doing the clerical work necessary to accomplish a change. Reducing the clerical time involved also can be done by making it easy for users to move quickly among several different tasks. When working on a program it is frequently necessary to browse other files in the system, to examine (or to send) electronic mail about the change in question, or to make use of one of the many network services that are available in a distributed environment. The costs of doing this switching should be minimal — it should not be necessary to "cleanup" and then "swap" to a new program.

The design of Cedar placed a good deal of emphasis on the ability to switch quickly among several different tasks. The underlying Cedar system allows a large number of processes to be efficiently created; each application is structured as one or more independent processes, so that long-running operations need not prevent further actions. And the applications themselves are written to be as unobtrusive as possible — they never preempt the user and force him to interact with only a single tool. The paper by Teitelman (30) shows many examples of the benefits of moving quickly among several different applications in Cedar.

1.3.3 Integrated Document/Program Composition System

Recently a number of "integrated environments" (Teitelbaum (28)) have incorporated syntax-directed editors; especially for novice programmers, these editors reduce some of the cost of getting a syntactically correct program produced.

In our environment (and most production environments), people spend much more time reading programs than writing them. Thus, improving the appearance of a program's text to make it easier to understand is more important than making it a little easier to write. So, instead of a syntax-directed editor, Cedar includes a complete document preparation system, with an (almost) WYSIWYG editor. (This paper itself was done using the Cedar editor, called *Tioga*.) Programs in Cedar are constructed using the same attention to formatting given to the papers we write —

comments appear in italics, section headings are given in bold, and indentation is used to show the structure of the program. (See Knuth's article (11) on "Literate Programming" for similar sentiments about the importance of treating programs like documents.)

Indeed, Tioga makes the on-line reading of programs much preferred over reading hardcopy. Having efficient string searches, being able to "level-clip" (showing only the high-level structure, rather than low-level details), and not having to flip pages makes it much easier most of the time to read a program on the screen rather than looking at a printed version. I generally do not print copies of programs I am working on, unless I am taking copies home for reading at night or on the rare occasion when I need to restructure so many interfaces that the display seems too small for all the pieces that I need to see.

1.4 MAKING CHANGES SAFELY

Most Cedar programmers spend a large portion of their time changing existing programs, rather than writing new ones: aside from the obvious necessity to fix bugs (either problems that have previously gone undetected or problems caused by unexpected changes in the operating environment of a program), there is always a good deal of work going on in improving systems by incorporating new algorithms or adding new functionality. Finally, many new programs get started by taking pieces of several existing programs and putting these pieces together in a new fashion.

Because of this, an important consideration in the design of the Cedar environment (carried over from the experience with Mesa) was to make it easy to make "safe" changes — to *know* that a change only affected some small region of a large, working system. The parts of Cedar that make this possible are based on the following assumptions:

The code for the system is the truth about the system. Comments in programs or other documentation may be helpful in understanding a program, but they can not be completely trusted. We've all seen cases where the documentation is outdated or mistaken.

Consistency checks are a necessity, not a luxury. It is important to catch errors, rather than having them remain undetected until long after they have occurred.

Focusing on the code of the system as the truth means that the programming language should allow the important assumptions upon which a program is based to be stated in the text (*not* as comments). Static typing is the most obvious example of this; the separation of Cedar programs into interfaces and implementations and the notion of a safe language subset in Cedar are additional variations on the same theme — these are all discussed below.

It is important to note that the same information used to determine the consistency of the system can also be used to help automate some of the tedious, error-prone tasks in managing a large system.   For instance, the Cedar system includes facilities to automatically perform reconstruction of a package after editing and to automatically retrieve or store consistent versions of all of the files that constitute a package; both of these are discussed below.

1.4.1 Interfaces and Implementations

While Cedar is a large system, it is built out of many small components. Following Mesa, components in Cedar are described by a separate *interface definition*, that is distinct from the actual *implementation*. The Cedar system currently consists of approximately 5000 program modules, divided roughly equally among definitions and implementations.

The benefits of making this distinction have been discussed several places and will only be briefly described here; they include:

1. An ability to work on the implementation and on client programs separately. While one obvious benefit of this is that these can proceed simultaneously, it also means that later changes to one does not necessitate changing (or even examining) the other. This is particularly important because it makes it possible for to make substantial changes to even fairly low-level implementations with confidence that the system will continue to hold together.

2. It is much easier in Cedar (and Mesa) to share code rather than to copy or write new code that duplicates previous system components. If a package provides the proper abstraction and acceptable performance, then no further details about its implementation

are necessary to use it. Note that this also prevents bugs from lingering longer than necessary in the system; sharing means that more clients may be bitten by a bug, but once it's fixed, there's no concern that large numbers of people may have copied the code and will be unaware of the repair.

3. Cedar programmers quickly learn to think in terms of interfaces, not implementations. In a system as large as Cedar, it is important to be able to understand the intended effects of many of the components in the system without having to read all of the code. Interfaces in Cedar and Mesa have proven to be a particularly attractive way to write specifications of the system — although we currently have no way of including formal specifications in an interface, it is relatively easy to write informal descriptions of an interface that capture the essential details in an understandable fashion.

One additional effect of the use of separate interfaces and implementations in Cedar is that everything in Cedar is better thought of as a package, rather than a tool — everything in the system has a programmer's, as well as a user's, interface. While programmer's interfaces to some components of the Cedar system are still afterthoughts, when done right they allow a very high degree of flexibility and integration. (Donahue(6)) gives some examples of the use of programmer's interfaces to high-level Cedar components.

The interface mechanism in the Cedar language is not perfect, however. The language syntax for importing interfaces is cumbersome (one generally has to write the same information three times in various guises). And having the compiler enforce interface boundaries also places a premium on careful design and documentation of the abstraction provided by the interface. While there are ways in Cedar to break into an implementation when the interface hides necessary functionality, this sort of thing is obviously discouraged. Finally, the current implementation of Cedar treats the changing of a comment in an interface as a change in the interface structure. Because of this, correcting an erroneous comment in a low level interface can be quite expensive, because of all of the recompilation that it might cause. This problem and the problem of the cumbersome syntactic conventions currently used in Cedar are relatively easy to fix; the need to do careful design before freezing interfaces, though, is a basic property of the Cedar philosophy. (Many consider it an important feature of Cedar, since it places emphasis on the design of *general* interfaces.)

1.4.2. Static Typing

Static typing is generally regarded as a desirable feature in a programming language because it allows the detection of an important class of programming blunders when a new program is being written. When we think about programs more in terms of long-term modification rather than initial composition, static typing takes on an even more important role — it ensures that programs that were constructed using assumptions that have changed will stop working.

Cedar, like Mesa, enforces strict type-checking of programs and version-checking of all of the interfaces used in a package; thus, the compiler (or the "binder", or linkage editor) will refuse to compile or bind a package that is type-incorrect or makes references to inconsistent versions of an interface. Interestingly, the static typing of Cedar frequently allows implementors to use a very simple procedure to rebuild a system:

1. Make the desired changes. This may involve changing the functionality of one or more procedures in a collection of interfaces.

2. Recompile the system. If the effects of the proposed changes are explicit in changes in the types of procedures, then any incompatible use will be caught by the compiler. Change all of the programs that won't compile any more. When the system is finally rebuilt, it should work.

In a system without strong type-checking, it is frequently necessary to debug a system back into existence following a change — there is no other easy way to determine whether all of the places dependent on the change have been successfully modified. Lauer and Satterthwaite (14) state that "The experience of many projects in Mesa is that once a previously running system has been successfully compiled and rebound following changes to its internal or external interfaces, it will immediately run with the same reliability as before."

One important effect of being able to isolate the effects of change is again to increase the ambition of implementors of large systems. If the system has to be debugged back into existence after making a change, then an individual or a small group may be quite reluctant to make modifications to the structure of a system — the cost of debugging the system (and the uncertainty that the last bug will be found) may be too high a price to pay for the value of the modifications. This tends to stifle the sort of experimentation that research environments are intended to do. In the development of Cedar, one can point to many instances of making very substantial changes to large parts of the system (even by people other than the original implementors).

1.4.3. The Safe Language

The Cedar system includes a garbage collector; the experience with the Mesa, Interlisp (Teitelman and Masinter (29)) and Smalltalk (Goldberg (8)) environments showed the potential benefits of not having client programs be completely responsible for storage management. However, the Cedar language is also intended to be used for even low-level systems programming; the Cedar garbage collector is itself written in Cedar. Because of this, the Cedar language contains constructs that could destroy the garbage collector invariants (the experience with Cedar has also shown us that there are some instances where acceptable performance can be obtained only when client programs do some storage management). And if the garbage collector should fail, it is important to be able to provide alibis for most of the system code (especially client programs using the allocator/collector in perfectly benign ways).

To do this, the Cedar language has a sublanguage, the "Safe Language", which is recognized by the compiler and in which it is not possible to write a program that violates the collector invariants. Most Cedar programs are written in the Safe Language and thus most Cedar programmers know that their programs can do no harm to the underlying storage system. Moreover, the Safe Language indicates what invariants need to be preserved when writing code that must be unsafe; these invariants are described in Owicki (17).

The safe language subset of Cedar is another example of the use of the programming language as the sole source of the truth about a program. If a failure of the storage abstraction occurs because of a violation of the collector invariants (a particularly hard class of errors to sort out), it is important to be able to find the possible sources of the error, while removing from consideration most of the program. The safe language gives sufficient conditions for deciding that some sections of a program could not possibly have been at fault: they are written in the safe language subset and they were compiled successfully. Since most of Cedar is now written in this safe subset, this means that most of the code can be discarded from consideration (with certainty that the problem won't later be found in a seemingly safe code segment).

This, at least, is the theory behind the safe language. Careful examination has shown that there are indeed some instances where programs certified as safe by the compiler can violate the storage invariants (the compiler does not insert bounds-checking code in all of the necessary places and does not handle forward- or self-references in initializations expressions properly). We intend to patch the places where known problems exist; however, the need for some more formal and rigorous understanding of the constraints on the safe language is clear.

1.4.4. Tools to reduce error

Performing the strong-type checking across module boundaries that is done in Cedar requires that each object file produced by the compiler includes version stamp numbers of each of the interface modules upon which the compilation depends (i.e., each of the interfaces imported or exported by the program). If importers and exporters of an interface disagree about the version, then the type-checking done by the system will fail (this can occur either when a package is being bound or when a package is loaded into running system). These same version stamps can be used to drive tools that automate many of the tedious, error-prone tasks in system construction.

1.4.4.1 Compilation. In constructing a large Cedar package, one must be aware of compilation dependencies (which modules import and export interfaces) and ensure that compilations are done in the right order. Getting this right the first time usually requires a little trial-and-error. However, once a package is built, it is possible to examine the object file for the package to determine how it can be rebuilt after some editing of the source files that make up the package.

Both the Mesa and Cedar environments include such facilities; the Cedar one is called *MakeDo*. After editing the sources for a package Foo, *MakeDo Foo* will do whatever compiles and binds are necessary to produce an up-to-date version of the package. The increase in productivity that such tools allow is hard to overestimate. The use of MakeDo means that a Cedar programmer has a simple means of guaranteeing that when he has completed his work, he will have produced a consistent version of a package. This makes it much more likely that small changes will be made to large packages by people other than the initial implementors.

1.4.4.2 <u>File storage and retrieval.</u> The same sorts of techniques applied to the problems of performing consistent compilation can also be applied to the problem of guaranteeing that a consistent version of a package is stored on a file server after being changed by an implementor. Cedar (and Mesa) use "description files" (called *DF files*) that contain a list of all of the files necessary to construct a package. The Cedar environment includes programs to retrieve all of the files referenced in a DF file from a file server and to copy back onto a server all of the changed components of a particular DF file. Finally, and most importantly, there is a program to check that the contents of a DF file are in fact a consistent and complete characterization of all of the source and object files necessary to make up a package.

The facility provided by DF files is particularly important in an environment for experimental programming, because it is common to have many versions of a package in existence. For instance, I currently am working on a new version of the electronic mail storage and retrieval system in Cedar. Because mail is a heavily used part of our environment, it is important that users be protected from test versions that are not ready for wide release. We currently have three active versions of the system: the one used by most Cedar users (which has been stable for several months), the one I am using (which tested out some of the new features, but was never intended for wide distribution) and the one that I am currently developing (which is not yet ready for daily use by anybody). Being able to easily manage these parallel versions is critical to making progress — even though the files for all three versions happen to be stored on the same file server, some even in the same directories, the use of DF files makes it relatively easy to keep all the pieces logically separated.

1.5 CONCLUSIONS

The successes of Cedar are the culmination of over a decade of work in both the Computer Science Laboratory and the System Development Division of Xerox (which uses its own Mesa-based *Xerox Development Environment* (Sweet (24))); in the references, I have tried to capture the important threads of this work — networked workstations, the Mesa language and processor architecture, open operating system architecture, and high-quality user interfaces. (The paper in this conference by Deutsch (5) discusses another important research effort in programming environments done at the Palo Alto Research Center.)

The Cedar system has proven to be an extremely useful tool in experimental programming. We do not have hard evidence on the performance gains that it has provided. However, our initial experience building experimental systems in Cedar (e.g., the Alpine file server, the Cypress database system, two different mail storage systems and a collection of VLSI design tools) has shown Cedar to have the advantages for experimental programming that were hoped for when the design began. We look forward to a great deal of work in the future, both in developing the Cedar system itself and in using it to build some unique applications software.

REFERENCES

1.   Beach, R. J., June 1984, 'Experience with the Cedar programming environment for computer graphics research,' Graphics Interface 84, Canada.

2.   Birrell, A. D., Levin, R., Needham, R. M., and Schroeder, M. D., April 1982, 'Grapevine: an exercise in distributed computing.' <u>CACM</u> <u>25</u>, 4, 260–274.

3.   Brown, M., Kolling, K., and Taft, E., 1984, 'The Alpine file system,' Xerox Palo Alto Research Center Report CSL-84-4.

4.   Deutsch, L. P and Taft, E. A., 1980, 'Requirements for an experimental programming

environment,' Xerox Palo Alto Research Center Report CSL-80-10.

5. Deutsch, L. P., 1985, 'Project support in the Smalltalk-80 integrated environment,' this proceedings.

6. Donahue, J. E., 1985, 'Integration mechanisms in Cedar.' ACM SIGPLAN Conference on Language Issues in Programming Environments, Seattle.

7. Geschke, C. M., Morris, J. H., Jr., Satterthwaite, E. H., August 1977, 'Early experience with Mesa,' CACM 20, 8, 540−553.

8. Goldberg, A. and Robson, D., 1983, 'Smalltalk-80: the language and its implementation,' Addison-Wesley, Reading, MA.

9. Harslem, E. and Nelson, L. E., 1982, 'A retrospective on the development of Star,' Proceedings of the Sixth International Conference on Software Engineering, Tokyo.

10. Johnsson, R. K. and Wick, J. D., March 1982, 'An overview of the Mesa processor architecture,' Proceedings of the Symposium on Architectural Support for Programming Languages and Operating Systems, Palo Alto.

11. Knuth, D. E., May 1984, 'Literate programming,' Computer Journal, 97−111.

12. Lampson. B. W., Mitchell, J. G., and Satterthwaite, E. H., 1974, 'On the transfer of control between contexts,' Lecture Notes in Computer Science 19, Springer-Verlag.

13. Lampson. B. W. and Redell, D. D., February 1980, 'Experience with processes and monitors in Mesa,' CACM 23, 2, 105−117.

14. Lauer, H. C. and Satterthwaite, E. H., September 1979, 'The impact of Mesa on system design,' Proceedings of the Fourth International Conference on Software Engineering, Munich.

15. Lauer, H. C., December 1981, 'Observations on the development of an operating system,' Proceedings of the Eighth ACM Symposium on Operating Systems, Asilomar.

16. Mitchell, J. G., Maybury, W., and Sweet, R. E., 1979, 'Mesa language manual,' Xerox Palo Alto Research Center Report CSL-79-3.

17. Owicki, S., 1981, 'Making the world safe for garbage collection,' Conference Record of the Eighth ACM Symposium on Principles of Programming Languages, Williamsburg, 77−86.

18. Pier, K., 1983, 'A retrospective on the Dorado, a high performance personal computer,' Xerox Palo Alto Research Center Report ISL-83-1.

19. Redell, D. D., Dalal, Y. K., Horsley, T. R., Lauer, H. C., Lynch, W. C., McJones, P. R., Murray, H. G., and Purcell, S. C., February 1980, 'Pilot: an operating system for a personal computer,' CACM 23, 2, 81−92.

20. Rovner, P., 1985, 'On adding garbage collection and runtime types to a strongly-typed, statically-checked concurrent language,' Xerox Palo Alto Research Center Report, to appear.

21. Shoch, J. F., Dalal, Y. K., Crane, R. C., and Redell, D. D., August 1982, 'Evolution of the Ethernet local computer network,' IEEE Computer 15, 8, 10−27.

22. Sproull, R. F. and Lampson, B. W., December 1979, 'An open operating system for a single-user machine,' Proceedings of the Seventh ACM Symposium on Operating Systems, Asilomar.

23. Sweet, R. E. and Sandman, J. G., Jr., March 1982, 'Empirical analysis of the Mesa instruction set,' Proceedings of the Symposium on Architectural Support for Programming Languages and Operating Systems, Palo Alto.

24. Sweet, R. E., 1985, 'The Mesa programming environment,' ACM SIGPLAN Conference on Language Issues in Programming Environments, Seattle.

25. Swinehart, D., Zellweger, P., and Hagmann, R., 1985, 'The structure of Cedar,' ACM SIGPLAN Conference on Language Issues in Programming Environments, Seattle.

26. Schmidt, E., 1982, 'Controlling large software development in a distributed environment,' Ph.D. Thesis, U.C. Berkeley EECS Dept. December 1982; also available as Xerox Palo Alto Research Center Report CSL-82-7.

27. Sheil, B., February, 1983, 'Environments for exploratory programming,' Datamation 29, 2,

131–144.

28. Teitelbaum, T. and Reps, T., September 1981, 'The Cornell program synthesizer,' <u>CACM</u> <u>24</u>, 9, 563–573.

29. Teitelman, W. and Masinter, L., April 1981, 'The Interlisp programming experience,' <u>IEEE Computer</u> <u>14</u>, 4, 25–33.

30. Teitelman, W., 1984, 'The Cedar programming environment: a midterm report and examination,' Xerox Palo Alto Research Center Report CSL-83-11.

31. Teitelman, W., April 1984, A tour through Cedar,' <u>IEEE Software</u> <u>1</u>, 2, 44–73.

32. Thacker, C. P., McCreight, E. M., Lampson, B. W., Sproull, R. F., Boggs, D. R., 1979, 'Alto: a personal computer,' Xerox Palo Alto Research Center Report CSL-79-11.

DSEE: Overview and configuration management

D.B. Leblang and G. McLean

ABSTRACT

DSEE (TM), the DOMAIN (R) Software Engineering Environment, supports large-scale distributed software development efforts on Apollo (R) workstations. DSEE provides source code control, task management, user-defined dependency tracking, and configuration management. An earlier paper (Leblang and Chase (7)) focused on the facilities offered by DSEE in the first three of these areas. This paper touches on all four areas, but the emphasis is on configuration management. The discussion treats not only the facilities offered by DSEE but also some of the technical issues involved in their implementation.

2.1 INTRODUCTION TO DSEE

DSEE supports large-scale distributed software development efforts involving engineers, technical writers, managers, and field support personnel. DSEE run on nodes in an Apollo network. The Apollo DOMAIN architecture provides a network-wide virtual address space, transparent remote file access, and remote paging (Leach et al (6)). DSEE uses a distributed database management system to store historical information; reliable, immutable, files to store deltas and tasks; server processes to watch for asynchronous events; and a store and forward inter-process communication mechanism to guarantee message delivery in case the network is temporarily partitioned. The DOMAIN system supports multiple windows, each of which may represent a seperate process. Some windows provide a general-purpose shell for executing system commands. Others are dedicated to applications such as mail. DSEE runs in a dedicated window that provides commands for activities directly related to software development.

DSEE consists of four managers:

o   A <u>History Manager</u> that provides source code control by maintaining historical source versions along with their histories.

o   A <u>Task Manager</u> that tracks source changes made anywhere in the network as part of some high-level activity.

o   A <u>Monitor Manager</u> that monitors user-defined dependencies and alerts the appropriate users when they are triggered.

o   A <u>Configuration Manager</u> that builds programs from desired source versions, detects the need to re-build programs when sources are updated, maintains a record of the source versions that were used in released configurations, and re-builds old configurations using bug fix versions of same sources.

2.2 HISTORY MANAGEMENT

The DSEE History Manager (HM) provides source code control. The unit of control is the <u>element</u>, which in a program development setting usually corresponds to a single source module or compilation unit. Related elements are grouped into <u>libraries</u>. A user <u>reserves</u> an element in order to make changes to it. The HM provides the user with a private copy of the element to work on. When he is satisfied with his changes, the user <u>replaces</u> the element, at which time the HM creates a new version of the element from the user's private copy. Whenever an element is reserved or replaced, the HM asks the user to describe the changes he intends to make or has made. The HM saves the user's response, together with the user's name and network location, and the date and time, in the library's <u>history database</u>. Thus, at any point in time users can determine which elements have been reserved, by whom, and for what reason. Users can also retrieve the history of an element, which essentially consists of the sequence of replace operations by which successive versions of the element were created. An especially useful facility provided by the HM is the ability to review the interleaved history for an entire library since a particular date, for example the date of an earlier major release. The HM also provides a sophisticated comparison facility by which different versions of an element can be compared in order to determine <u>exactly</u> what changes were made.

The HM assigns a <u>version number</u> to each successive version of an element at the time the version is created. This version number, which is a small integer, distinguishes the version from all other versions of the same element. A user can refer to a particular element version by means of its version number, which the HM displays whenever it prints information about the version, for example when it shows the history for the element. Version numbers have no mnemonic value, however, and they do not provide a way for relating versions of a whole set of elements. The HM therefore allows the user to assign a <u>version name</u> to any version of an element. Typically, version names are used to capture a cross section of element versions for an entire library. For example, if the library contains the source modules for a program, then at the time of a major release of the program, say Release 9.0, the user might instruct the HM to name the most recent version of every element in the library "release_9.0". (This can be done with a single HM command.) Thereafter, the user can refer to the Release 9.0 version of an element using the version name instead of the version number (e.g. "gpr.pas[release_9.0]" instead of "gpr.pas[42]", where "gpr.pas" is the name of the element). In general, version names may be used in any DSEE command that takes an element version. Despite its simplicity, this has proven to be a surprisingly powerful way of tracking and manipulating released configurations. More will be said about this later.

An element always has a <u>principal line of descent</u>, which consists of a succession of "mainline" versions in the order they were created. By default the reserve and replace commands serve to extend this line of descent. However, an element may also have one or more named <u>variant lines of descent</u>, or <u>branches</u>, which also consist of a chronological succession of versions and represent competing lines of development. Different branches of an element can be reserved and replaced independently, thereby extending the lines of descent they represent. A user might create a branch off of an old mainline version in order to fix a bug in that version without incorporating more recent changes to the element. Alternatively, at the start of an extended period of parallel development, a user might create branches with the same name off of the most recent versions of a whole set of elements. Finally, a user might create a temporary branch in order to work on an element that is currently reserved by someone else. The HM provides an interactive merge facility that can be used to re-unite two lines of descent by merging their most recent versions. The merge operation produces the next version of the on-going line of descent (typically the principal line of descent), and "terminates" the obsolescent line of descent.

DSEE elements appear under their user-specified names in the file system's naming hierarchy. That is, DSEE elements may be referred to by pathname outside the DSEE environment just the way other files are. More importantly, historical versions of an element may be obtained without "fetching out" copies into temporary files. The most recent mainline version of an element is always available by means of the normal device-independent stream I/O interface supplied by the operating system. By default, whenever a program opens an input stream on an element, it "sees" the most recent version of that element, just as though the element were a normal text file. Any translator, formatter, or other program, be it Apollo-supplied or user-written, that operates on text files using the standard stream I/O interface can therefore operate on DSEE elements. Moreover, when other than most recent versions are desired, DSEE may be used to set up a per-process <u>version map</u>, which specifies on a per-element basis what versions should be materialized when programs running in the process read those elements using the stream I/O interface.

It has been our experience that this transparent access to elements is crucial to user acceptance of the source code control system. It is made possible because parts of the HM are implemented within the Apollo operating system.

An element's versions are stored in a single file in a highly compressed format that uses "interleaved" incremental deltas. This format makes it possible to reconstruct efficiently and incrementally any version of the element during a single sequential pass over the file, which is required in order to provide transparent record-at-a-time access, as described above. In addition, the space savings gained by storing deltas instead of whole

versions can be impressive. It is not unusual to be able
to represent 50 or more versions of a source element in no
more space than would normally be required to hold two full
copies of the element.

The "meta-data" for a library, including element
histories and other control information, are stored in a
database associated with the library. The database is
independently controlled by the Apollo distributed database
management system, which supplies the sophisticated data
structuring facilities needed to represent the library's
meta-data and also guarantees its recoverability in the
event that problems occur while it is being updated.   DSEE
coordinates updates to the library database with updates to
element files in such a way that user-level operations
appear to be executed atomically: they either succeed
entirely or leave no trace in the library.

2.3 TASK MANAGEMENT

The DSEE Task Manager (TM) provides a way to plan and
track the low-level steps involved in some high-level
activity. Such high-level activities, or tasks, include
making a specific enhancement to a program or fixing a
particular bug, which often entails changes to a number of
elements, perhaps in several libraries. The TM can
automatically maintain a record of these changes, thereby
making it easy to determine at a later time exactly what
was done as part of the task.

A DSEE _task_ consists of a user-supplied title and list
of _active items_, together with a _transcript_ of _completed
items_. Active items represent anticipated steps that are
yet to be taken, and together form a plan of action.
Completed items represent steps that have been taken, and
include formerly active items that the user has "checked
off", together with items added automatically by the TM.

In order to work on a task, the user makes it his
current task. Thereafter, whenever the user replaces an
element in any library, the TM records the replace
operation in the task transcript in the form of a completed
item. Like the library history entry created by the
replace operation, the completed item contains the name and
network location of the user, the date and time, the name
of the element that was replaced and the version that was
created, and the user's description of the change. The
same task may simultaneously be the current task for more
than one user. In this case the task transcript reflects

the activity of all such users.

Tasks are organized using <u>tasklists</u>. By default each user has a <u>personal tasklist</u>, and each library has a <u>master tasklist</u>. The user may also create other tasklists as the need arises. The same task may appear on more than one tasklist. For example, when a user creates a task, the TM adds references to it to the master tasklist for the current library, and to the user's personal tasklist. The user may elect to add the task to the personal tasklists of other users as well, if more than one user will work on the task. When a user has finished working on a task, he normally instructs the TM to "drop" the task from his personal tasklist. A task usually remains catalogued indefinitely on the master tasklist, however, since it contains a valuable record of what was done as part of the task.

Tasks and tasklists provide a useful tool for tracking bug fixes. Before setting out to fix a bug in a released program, the user might create a task whose title contains the bug report number. After selecting the task as his current task, the user would proceed to reserve and replace elements in various libraries in order to make the necessary changes. These operations would automatically be recorded in the task transcript. When he was satisified that the bug had been fixed, the user would drop the task from his personal tasklist, and add it to a special bug fix tasklist. At a later time the bug fix tasklist could be examined to determine what bugs had been fixed, and tasks on the tasklist could be individually examined to determine exactly what changes had been made.

Tasks can also be useful when far-reaching changes or enhancements are made to a program. In this case the task transcripts automatically maintained by DSEE can help users keep track of the changes they have made, and may ultimately serve as the basis for a code review or even help in "backing out" some or all of the changes if a better implementation is found.

2.4 MONITOR MANAGEMENT

The DSEE Monitor Manager (MM) provides a way for users to be notified when particular elements are changed. To see why such notification may be important, consider the relationship between a program's software and its user documentation. When the source module that implements the program's user interface is changed, it is likely that help files and other forms of user documentation will have to be updated, so the technical writers need to be made aware of the change. The MM provides a flexible mechanism for automatic notification that can be used to satisfy this requirement.

A monitor consists of a title that describes its purpose, a list of the elements to be monitored, a task "template" to be "instantiated" when the monitor is activated, a list of the tasklists to which the instantiated task should be added, and a list of the shell commands that should be executed. When a user reserves a monitored element, he is informed of.that fact and is shown the monitor's title and the name of the person who created it. In this way the user is made aware of the "dependency" that the monitor represents before he begins to modify the element. When a user replaces a monitored element, the monitor is activated. This involves instantiating the monitor's task template to create a new task that is then added to the tasklists specified by the monitor. If the monitor contains shell commands, they are also executed at this time.

Typically a monitor's task template contains a list of active items that represent the steps that should be taken when the monitor is activated. The monitor's tasklists are usually the personal tasklists of the people on whom the responsibility for taking those steps falls. A user can arrange to have a small alarm window automatically pop up on his screen whenever monitor activation adds a task to his personal tasklist.

At the time a monitor is created, the MM allows the user to specify that the monitor should be activated only when he replaces the element, or, alternatively, only when someone else replaces the element. Regular expressions may be used to specify the elements to be monitored, so it is a simple matter to create a single monitor that monitors all elements in a library.

2.5 CONFIGURATION MANAGEMENT OVERVIEW

The DSEE Configuration Manager (CM) builds programs from desired element versions and maintains a record of the element versions that were used in released configurations. When building a program with the CM, the user specifies only the desired element versions and translation options. The CM handles finding any existing binaries that satisfy the user's specifications, executing

the appropriate translation rules to derive the remaining
binaries, and associating a record of the versions used
during the build with the binaries it produced. The CM
thus assumes responsibility for managing binaries.
However, it is a major objective of the CM that it have no
special knowledge of the translators that produce those
binaries. As a result, user- as well as Apollo-supplied
translators may be employed with equal ease, thereby making
the CM language-independent as well as target
machine-independent.

The CM makes it possible to

o   capture and enforce the "protocol" for correctly
    re-building a program after changing one or more of its
    constituent elements

o   specify in a high-level way the element versions and
    translation options to be used in building a program

o   build different configurations of a program, possibly
    for different target machines, without copying sources
    or binaries

o   share binaries wherever possible among users working on
    different configurations of a program

o   establish, without copying sources or binaries, a
    stable environment in which a user is isolated from the
    changes of others and "sees" only the changes he makes

o   create a permanent record of the element versions used
    in a released configuration of a program

o   use the permanent record of constituent versions for a
    released configuration to establish a process
    environment in which programs that access DSEE elements
    automatically "see" the element versions that were used
    in that configuration

o   re-build an old configuration of a program using bug
    fix versions of some of its constituent elements

The following sections discuss the central concepts
introduced by the CM.

2.5.1 System Models.

Because the CM can be used to build not just programs but suites of programs, software libraries, user manuals, and so forth, the more general term _system_ is used to refer to the unit of configuration manangement. A system is defined by a _system model_ (Lampson and Schmidt (9), Teitelman (13)). The system model gives a static description of the structure of the system in terms of its buildable _components_, their constituent elements, and their hierarchical relationships. The system model also gives the _translation rules_ to be used to produce the _derived objects_ for buildable components. Translation rules can include _translation options_, which may or may not be requested by a user during any given build of the system.

For each buildable component of the system, the system model declares the name of the component, the elements and buildable components on which it depends, and the translation rule to be used to produce its derived object(s). The translation rule is essentially a stylized "shell script" that contains the compiler, binder, or formatter command lines needed to build the component. The declared dependencies of buildable components on other buildable components impose a hierarchical relation on the set of components that ultimately affects the order in which they are built. Roughly, if the system model declares that component X depends on component Y, and both X and Y must be built in order to build the system, then Y will be built before X.

The system model is "source-oriented". Instead of describing the structure of a system in terms of its derived objects (Feldman (3)), it focuses on what buildable components there are and what source elements comprise them. Even translation rules do not refer directly to derived objects. Rather, they speak indirectly of "the result of translating" certain components.

The system model is written in a block-structured _system model language_. The block structure mirrors the hierarchical relationship among buildable components, and also provides scoping for certain kinds of declarations. Special constructs are provided for indicating where in a translation rule the CM should substitute the pathnames of the relevant derived objects. Such embedded constructs are necessary for two reasons. First, only the CM knows the pathnames of derived objects, and even then only at build time, not compile time. Second, only the user knows where those pathnames should ultimately appear in any shell script derived from the translation rule. (Remember that the CM has no special knowledge of the translators used to build a system). Similar constructs are provided for indicating where translation options specified by the user at build time should be substituted. Aside from these special embedded constructs, the CM treats a translation rule as uninterpreted text.

The system model language allows declarations that are common to a set of buildable components to be "factored out". For example, if every buildable sub-component of component X depends on a certain element (an "include file"), the system model can state this succintly with a declaration at the block level for X, which also defines the scope of the declaration. The set of components to which a factored declaration applies may be further restricted by means of a regular expression for the component names. It is therefore a simple matter to specify a single translation rule to be used for every buildable component whose name ends with a certain extension (e.g. ".pas" or ".c").

A conditional compilation facility makes it possible to parameterize a system model in a way that allows it to be used to build structurally variant configurations of the same system, e.g. configurations for different target machines. Conditional compilation is controlled by "switches" supplied by the user when he makes the system model his current system model.

2.5.2 Configuration Threads.

The system model gives a static description of the structure of a system. It does not specify what versions to use for its constituent elements during any given build. Specifying desired element versions is the function of the user's <u>configuration thread</u> (CT). A CT consists of an ordered list of rules. Each rule contains a predicate that identifies the elements to which it applies, and a specification of the version to use for those elements. The predicate may stipulate that the rule applies only to the element with a given name, or more generally to elements whose names match a given regular expression. The predicate may go on to stipulate that the rule applies only to elements that are currently reserved by the user from the source code control system, or alternatively only to elements having a branch with a given name. The version specification may be static ("use version 42 of the element", "use the version named 'release_9.0'") or dynamic ("use my working copy of the reserved element", "use the most recent version of the element").

The order in which rules are listed in the CT is significant. To determine what version to use for a given element, the CM selects the first rule whose predicate is satisfied by the element. CT rules are therefore listed in order of increasing applicability of their predicates. A couple of examples should illustrate this point. A developer working on the next release of a system would typically want to use the most recent versions of all elements, except that he would want to use his working copies of any elements that he currently had reserved. His CT might state

1. If the element is reserved then use my working copy.
2. Otherwise use the most recent version of the element.

Someone working on a bug fix to a past release of the system would probably also want to use his working copies of any elements that he currently had reserved. However, instead of the most recent versions of other elements, he would want to use the same versions that went into the release, unless he had created a branch off of one or more of those old versions in order to fix the bug, in which case he would want to use the branch versions. His CT might state

1. If the element is reserved then use my working copy.
2. If the element has a branch named "release_9.0_bugfixes" then use the most recent version of that branch
3. Otherwise use the version named "release_9.0"

This CT assumes that all element versions that were used to build the system for Release 9.0 have been given the name "release_9.0" (using the HM). Actually, a CT rule can also call for the versions used in an earlier release simply by referencing the release's permanent record of constituent versions. We will return to this point later.

To simplify the discussion we have assumed that CT's are "flat", i.e. that for each element they specify a single version to be used whenever the element is needed during a build of the system. In fact, CT's may specify

that different versions of an element should be used in
different contexts. For example, if buildable components X
and Y both depend on the element Z (an "include file"), the
CT can specify that one version of Z should be used when X
is built and a different version of Z should be used when Y
is built. This facility makes it possible to use a bug fix
version of an element in exactly those contexts that need
the bug fix, leaving the remaining contexts to use the
original element version.

    We close this section by mentioning that CT's may also
specify translation options for some or all buildable
components.   Thus, if the user wanted "debug-mode" binaries
for all reserved source modules, his CT might include a
rule that states "If the component depends on a reserved
element then use (i.e. compile with) the '-debug' option".

2.5.3 Bound Configuration Threads.

    The user's CT rarely specifies an explicit version for
every element in the system. Instead it uses constructs
like "the most recent version of the element". The CT must
therefore be evaluated at build time in order to bind each
element in the system to a desired version. The result of
doing so is a bound configuration thread (BCT). A BCT is
by nature hierarchical, since different versions of the
same element may be called for in different contexts. In
addition to specifying explicit versions for all elements
in the system, a BCT also specifies translation rules for
each of its buildable components, including any translation
options called for by the CT. A BCT is thus a complete
specification for building a configuration of the system.
It also serves as a valuable record of what went into such
a configuration, as we shall see.

2.5.4 Derived Object Pools.

    Associated with each system are one or more derived
object pools, in which the binaries and other derived
objects for buildable components are normally kept.
Different systems can share the same derived object pools.
For example, the system for a Pascal compiler might use two
pools, one to hold the derived objects for its front-end
components and another to hold the derived objects for its
back-end components.   The same back-end pool might also be
used by the system for a C compiler, assuming that the two
compilers have a common back-end.

In order to re-use derived objects produced by earlier builds, the CM must be able to tell what versions and translation options were used to produce each derived object in the pool. One approach to solving this problem would be to store that information in the derived objects themselves.   This,   however,   would   require   special cooperation between the CM and translators, and would therfore   effectively   restrict   use   of   the   CM   to Apollo-supplied translators.   For this reason the CM associates with each derived object an auxilliary object in which it stores the derived object's BCT. As we pointed earlier, the BCT provides an exact specification of the versions and translation options that were used to produce the derived object.

At any given time the derived object pool may contain several derived objects for the same buildable component. This results from the fact that developers working on different configurations of the system use the same pool, in order to maximize sharing of the derived objects they have in common.   The CM deletes derived objects from the pool as they fall into disuse, in accordance with a user-specified limit on the number of derived objects per component. If inserting a new derived object for a given component would exceed this per-component limit, the CM first deletes another of the component's derived objects, using a least-recently-used (LRU) replacement algorithm. Should the deleted derived object be needed again in the future, the CM can re-derive it from its constituent element versions, which are always available from the source code control system.

Derived objects built from working copies of reserved elements are of no interest to anyone but the user who has those elements reserved, so it makes little sense to place them in a central pool devoted to sharing.   Moreover, such derived objects cease to be of interest even to the user who built them as soon as he edits his working copies and builds new derived objects. Thus they ought to be deleted more quickly than derived objects in the central pool would be.   For these reasons the CM provides each user with a <u>reserved pool</u>, into which it places all derived objects built   from   the user's working copies.   By default a reserved pool's per-component limit on the number   of derived objects is one, so each time a new derived object is inserted into the reserved pool it replaces any existing derived object for the same component. A derived object is <u>promoted</u> from the user's reserved pool to the central pool

if the user replaces (into the source code control system)
all of the working copies from which the derived object was
built.

At an implementation level, the three fundamental
operations on a derived object pool are __insertion__, __lookup__,
and __validation__. Insertion creates a new derived object and
associated BCT in the pool, which may have the side effect
of causing another derived object for the same component to
be deleted. Lookup provides associative access to derived
objects in the pool based on the desired BCT. That is, it
finds a derived object whose BCT matches the desired BCT,
if such a derived object exists. Validation verifies that
a derived object found by an earlier lookup still exists in
the pool.

A hashing scheme is used to ensure good performance
for BCT lookup. As a result, lookup rarely requires more
than one pool BCT to be examined in order to find the
desired derived object, if it exists in the pool. If the
desired derived object does not exist in the pool, lookup
normally determines this fact without examining __any__ pool
BCT's.

2.5.5 Building.

Once the user has selected a system model and a
configuration thread, he can issue a command to build all
or part of the system. The CM first evaluates his CT in
order determine what element versions and translation
options to use. The result is the __desired BCT__. The
desired BCT specifies element versions and translation
rules for the component, its sub-components, their
sub-components, and so on.

Simply stated, the CM looks up the desired derived
objects in the derived object pool. Those that exist it
re-uses, those that do not it builds in accordance with the
desired BCT. Builds are performed in the order implied by
the system model. That is, if components X and Y must both
be built, and X depends on (uses) Y, then Y is built before
X.

Notice that in the preceding discussion of how the CM
determines what needs to be built, no mention is made of
elements having "changed". The CM does not build something
because some element or working copy has "changed", but
only because the desired BCT calls for a derived object
that does not currently exist in the pool. Of course, most
often the reason that the derived object doesn't already

exist in the pool is that the user has just created a new version of some element for which his CT requests the most recent version, or has edited his working copy of a reserved element. However, the CM does not distinguish this case from one in which the user is simply trying to re-build an old configuration whose original derived objects have long since dropped out of the pool due to old age.

Suppose that the CM has determined that it needs to build component X. Before it can do so, it must do two things. First, the CM must derive an _effective translation rule_ from the system model translation rule for X. This involves substituting into the translation rule any additional options requested by the user, and also the pathnames of derived objects, such as those for the sub-components of X (which the CM has already found in the pool or built). Second, the CM must establish a process environment in which the desired element versions are automatically materialized whenever elements are read by programs executing in that process. This is accomplished by setting up a process-specific version map in accordance with the desired BCT. Such version maps were discussed earlier in the section devoted to the History Manager.

A couple of examples will help to motivate the preceding discussion. Suppose that the derived object for X is produced by binding (linking) together the binaries for a number of lower-level components that correspond to source modules. We assume for the sake of this discussion that the binder program expects the pathnames of the binaries it is to bind together. Instead of specifying those pathnames (which are unknown at system model compile time), the system model translation rule for X will contain special constructs that call for "the result of translating" the lower-level components. At build time the CM has determined the pool pathnames of their binaries, so it can substitute those pathnames into the translation rule at the indicated positions. It can also substitute the pathname that will cause the binder to output the derived object for X directly into the pool.

Now suppose instead that X is a source module that must be compiled. In this case the system model translation rule for X will simply invoke the appropriate compiler, and the only derived object pathname that will need to be substituted into it is the one for the resulting binary. However, the CM must go on to ensure that when the compiler reads the source element for X, and the source elements for any "include files" used by X, it "sees" the desired versions. This it does by means of a process-specific version map.

Once it has computed the effective translation rule and established the desired version map, the CM can invoke a shell to execute the translation rule. The shell is normally run in a separate process at the user's node. However, the user may specify that builds are to be performed in a process located at some other node in the network, for example a "computation server" that has a more powerful CPU.

2.5.6 Not Building.

In a large system an element may be listed as a dependency for hundreds of components. This might be the case if the element were an "include file" containing global declarations needed by nearly every source module in the system. Changes to such an element would normally cause the CM to re-build most components in the system, if the CT calls for the most recent version of the element. This is frequently unnecessary, because the changes are such that only a small number of components are affected. For example, adding new declarations to an include file often qualifies as such a change.

In a development environment it is infeasible to re-build the entire system each time an include file is changed. Forcing users to edit their CT's when an include file changes is no solution, because a CT that explicitly requests an old include file version for some components blinds users to future changes to the include file. The CM therefore allows a user to declare that a new version of an element is <u>equivalent</u> to the previous version for certain components that depend on the element. The CM interprets this to mean that for those components, derived objects built from the new element version are interchangable with ones built from the previous version. Equivalences are normally declared by the user making the change the next time he builds the system. They are stored in the derived object pool in the form of BCT's that have no derived objects of their own but instead refer to the derived objects associated with other BCT's. If during pool lookup the CM finds an equivalence that matches the desired BCT, it uses the derived object to which the equivalence refers, as long as the user has indicated that equivalent derived objects are acceptable. (Of course, BCT's inserted into

the pool reflect the versions that were actually used).
Storing equivalences in the pool thus makes them available
to all users, most of whom are not in a position to
determine what components were affected by the original
user's changes.

The system model language also provides a way for
users to declare non-critical dependencies. Suppose that
component X is declared to have a non-critical dependency
on component Y. Then during pool lookup the CM will accept
any derived object for X whose BCT differs from the desired
BCT only with respect to Y. If the CM does decide that X
must be built, however, it will consult the user's CT to
determine what element versions to use for Y, as usual.
Thus, changes to elements on which there are only
non-critical dependencies are never sufficient to cause the
CM to re-build anything. Nevertheless, non-critical
dependencies differ from undeclared dependencies in that
BCT's still provide a record of the versions used for all
constituent elements, whether the dependencies on them are
critical or non-critical.

A development team might declare dependencies on
global include files to be non-critical, since they would
not normally want the CM to re-build all components when
new declarations are added to the include files. Normally
the source modules that are affected by the new
declarations must themselves be changed to make use of
them, so the CM would want to re-build those modules
anyway, at which time it would use the include file
versions called for by the CT. When an incompatible change
is made to a global include file, the user can force the CM
to re-build all components.

2.5.7 Optimizing Build Performance.

A straightfoward implementation of the algorithms
sketched in the preceding section could add a great deal of
overhead to every build, thereby making it prohibitively
expensive to use the CM for day-to-day development and
maintenance of large systems. For this reason a number
significant optimizations have been included in the
implementation.

Most configuration threads call for the most recent
versions of some or all elements in the system. Since a
system may be comprised of hundreds of elements, it is
infeasible to interrogate them all at the start of every
build in order to determine what their most recent versions

are. Nevertheless, users may create new versions of elements at any point in time and from any node in the network. To resolve this dilemma, the CM maintains a per-user cache of the most recent version numbers for elements in the system. It can take some time to initialize this cache the first time the user (ever) builds the system, if hundreds of elements are involved. Thereafter, however, the CM updates the cache incrementally as new element versions are created, and this goes very quickly. Discovering that a new element version has been created proceeds in two steps. Whenever a user replaces an element, the normal file system operations involved in doing so cause the date/time-modified (DTM) fields of the file containing the element, and of the library in which the element resides, to be updated by the operating system. By saving library DTM's in the cache, the CM can easily determine at the start of a build which if any of the libraries used by the system have been updated since the last build. By also saving element file DTM's along with most recent version numbers, the CM can then easily determine which elements in those libraries were updated. It must then re-interrogate those elements in order to update the cache.

Because of the central role it plays in optimizing build performance, a user's version cache is never discarded. Instead it is saved across sessions in a per-user area associated with the system.

It is also impractical to make every build compute a desired BCT for the entire system from scratch and then look up all of the requested derived objects in the pool. To avoid doing so the CM takes advantage of the fact that major parts of the desired BCT remain unchanged from build to build. That is, each new build will request the same versions for most elements as the previous build. The CM therefore maintains a <u>previous desired BCT</u> for each user, which is simply the desired BCT computed during the user's last build, augmented with the pathnames of the derived objects used in that build. At the start of a build, the CM quickly determines what elements have new versions, as described in preceding paragraphs, and also what working copies of reserved elements have been edited by the user. The CM then computes the new desired BCT incrementally by updating those portions of the previous desired BCT that reference the changed elements or working copies. The remaining, unchanged portions of the previous desired BCT are flagged as being "the same as previous". For those components with respect to which the new desired BCT is the same as the previous one, the CM simply re-uses the derived objects it used in previous build, after validating that they still exist in the pool. The remaining components must be looked up associatively in the pool.

Like the user's version cache, the previous desired BCT is saved across builds in a per-user area associated with the system.

2.5.8 Release Management.

Most of the builds performed during the normal course of development produce derived objects that are of fleeting utility. They quickly become uninteresting, and unused, as new element versions are created, and so rapidly disappear from the derived object pool, taking their BCT's with them. The situation is very different for derived objects built with the intention of releasing them to the outside world. In this case the user will want to have permanent copies of some or all of the derived objects and, just as importantly, a permanent record of the element versions and translation rules used to build them.

The DSEE Release Manager (RM) provides facilities for tracking and manipulating released configurations of programs. A _release_ consists of a name, some or all of the derived objects produced by a build, the BCT produced by the build, a snapshot of some or all of the source element versions and tools used in the build, and additional user-supplied information about the release. The user-supplied name and other information should be sufficient to identify the release for the user's purposes. Information about releases is kept in an area associated with the system. Users may copy releases to archival storage.

To see how the information saved in a release might be used, suppose that a bug is discovered in Release 9.0 of the system, which has been given the name "release_9.0". A user charged fixing the bug can examine the BCT for the release in order to determine what element versions were used in it. He can then use the History Manager to peruse those element versions. He will probably need to make changes to some elements in order to fix the bug. To do so he will create branches off of the element versions used in the release, and then proceed to reserve and replace those branches as he works on a bug fix. It will be particularly convenient to give all such branches the same name, say "release_9.0_bugfixes". To build a configuration of the system that is just like Release 9.0 except that it uses his bug fix versions, the user might employ a CT that reads

1. If the element is reserved then use my working copy.
2. If the element has a branch named "release_9.0_bugfixes" then use the most recent version of that branch
3. Otherwise use the version used in release "release_9.0"

Note that the last rule in the configuration thread refers to the BCT for the release.

Examining and manipulating the element versions used in a release is made much simpler if they are given the name of the release. (See the discussion of version names in the section devoted to the History Manager). The RM allows the user to specify that all element versions used in a release are to be given a particular name, typically the name of the release. It then becomes possible to refer to those versions throughout DSEE simply by giving the version name. Assuming this has been done in the current exmple, the user can see the released version of the element "gpr.pas" by asking the HM to put up a read-only display of "gpr.pas[release_9.0]". If necessary the user can then go on to ask the HM to create a branch off of "gpr.pas[release_9.0]". In general it is much more convenient to use such mnemonic version names than it is to use numeric version numbers culled from BCT's.

2.6 CURRENT STATUS

At the time of this writing, an initial version of the DSEE product that includes a full implementation of history, task, and monitor management has been in use in the field for over a year. DSEE has been used internally at Apollo for more than two years by diverse groups developing microcode, operating systems, graphics software, language translators, documentation, and DSEE itself. The product's reception has been gratifying.

The DSEE Configuration Manager is still under development. For the past six months DSEE developers have used a subset implementation of the CM in order to do further development work. This subset is now in beta test. Remaining to be implemented are release management and some of the more sophisticated features of configuration threads.

2.7 ACKNOWLEDGEMENTS

The authors wish to acknowledge the contributions made to this paper, and to DSEE itself, by DSEE project members Howard Spilke, Bob Chase, Pat Bergeron, John Yates, Don Curreri, Erica Dorenkamp, and Debbie Minard.

APOLLO and DOMAIN are registered trademarks of Apollo Computer Inc. DSEE is a trademark of Apollo Computer Inc.

REFERENCES

References 1, 2, 4, 5, 8, 10, 11, and 12 are not cited in the body of the paper.

1.   Source Code Control System User's Guide
      UNIX System III Programmer's Manual, Oct 1981.

2.   CMS/MMS: Code/Module Management System Manual
      Digital Equip. Corp., 1982.

3.   Feldman, S.I.
      "Make - A Program for Maintaining Computer Programs"
      Software Practice and Experience, Apr 1979.

4.   Heckel, P.
      "A Technique for Isolating Differences Between Files"
      CACM, Apr 1978.

5.   Ivie, E.L.
      "The Programmer's Workbench"
      CACM, Oct 1977.

6.   Leach, P., Levine, P., Dorous, B., Hamilton, J.,
      Nelson, D., Stumpf, B.
      "The Architecture of an Integrated Local Network"
      IEEE Journal on Selected Areas in Communications,
      Nov 1983.

7.   Leblang, D. B., Chase, R.,
      "Computer-Aided Software Engineering in a Distributed
      Workstation Environment"
      ACM/SIGPLAN/SIGSOFT Conference on Practical Software
      Development Environments, Apr 1984.

8. Lampson, B., Schmidt, E.
    "Organizing Software in a Distributed Environment"
    SIGPLAN Jun 1983.

9. Lampson, B., Schmidt, E.
    "Practical Use of a Polymorphic Applicative Language"
    10th POPL Conf., Jan 1983.

10. "STONEMAN: Requirements for Ada Programming Support
    Environments"
    U.S. Department of Defense, Feb 1980.

11. Thall, R.
    "Large—Scale Software Development with the Ada
    Language System"
    Proc. of ACM Computer Science Conf., Feb 1983.

12. Tichy, W. F.
    "Design, Implementation, and Evaluation of a Revision
    Control System"
    6th Int'l Conf on Software Eng., Sep 1982.

13. Teitelman, W.
    "Cedar: An interactive programming environment for a
    Compiler-Oriented Language"
    LANL/LLNL Conference on Work Stations
    in Support of Large Scale Computing, Mar 1983.

Chapter 3

Project support environments for formal methods

I.D. Cottam, C.B. Jones, T. Nipkow, A.C. Wills, M.I. Wolczko, A. Yaghi

The authors would like to acknowledge the financial support of the UK Science and Engineering
Research Council and ICL

There is a growing acceptance of the need for formal methods in
software development. The design of Project Support Environments should
recognize the special requirements that this will generate. The authors
have experimented with existing Syntax Directed Editors and are now
developing a Structure Editor (known as "Mule") to support formal
development methods such as "VDM": a progress report and some tentative
conclusions are given.

3.1 INTRODUCTION

This paper attempts to show how project support environments can be
used to ease the use of formal methods for program specification and
design. It is argued that a fine-grained and strongly interconnected
database is required.

The term *formal methods* covers formal specification and
verification. Formal specifications use mathematical notation to
achieve precision. Formal verification can be used to show that a
program satisfies a (formal) specification. Verification techniques can
be applied during the design process: the most productive time to
construct a justification of a design step is before further work is
based on it (cf. Jones (18)).

There are a number of different approaches to formal methods. The
algebraic (property oriented) approach is described in Ehrig and Mahr
(7). The development of programs by transformations is explained in
Bauer and Woesner (2) and significant examples are given in Partsch (26)
and Reif and Scherlis (27). The Vienna Development Method (VDM) is
covered by Jones (16) and Bjorner and Jones (4) with an account of some
recent developments in Jones (17). VDM describes data types by
constructing models and defines operations by pre- and post-conditions.
Proof obligations for data refinement and operation decomposition are
given. The main conclusions in this paper apply to support tools for
other model oriented techniques such as "Z" (cf. Hayes (15), Morgan and
Sufrin (24)) and many of the observations would also appear to be
relevant to other approaches to formal methods.

Floyd (8) attributes:
> "I would rather write programs to help me write programs
> than write programs".

to a Stanford graduate student. The use of VDM with nothing more than a
text-editor has shown a pressing need for programs which help with the
formal specification and systematic development of programs. A VDM

specification is a text in a formal language. At least initially, the syntax of this language is less familiar to a user than those of programming languages. It is, therefore, likely that the Syntax Directed Editors (SDEs) – such as are used in Gandalf (Medina-Mora and Notkin (22)), Mentor (Donzeau-Gouge et al (6)) or Cepage (Meyer and Nerson (23) – could make the use of VDM significantly easier. Appropriately designed tools can contribute significantly to the promulgation of good software engineering methods.

Design requires invention which is aided by intuition. The record of the design process presents an idealized picture leaving aside the many false starts. Using a formal approach, each level of design is written in a formal language and the resulting proof obligations can be mechanically generated. The degree of formality to be used in proofs depends on the reliance to be put on the finished program. Formal proofs are difficult to develop but it is easy to find ways in which mechanical aids can help with routine tasks like substitution and keeping track of the known facts and goals. Of course, the ultimate goal is for far more elaborate aids to theorem proving – but this goes beyond the scope of the current paper.

There is considerable experience with formal specifications of increasing size. One of the largest VDM specifications is that for the Ada programming language (cf. Bjorner and Oest (3)). The construction of such documents would benefit from tools. But the time when the tools become mandatory is when such documents have to be modified. The whole point of constructing formal specifications can be destroyed by changes which lead to inconsistencies.

There are a number of requirements on support environments for projects which employ formal methods. The requirements of any project support environment apply: thus it is necessary to provide:
 – a helpful User-System Interface (USI)
 – a single, integrated, database
 – support for version control.
The material below explains why the need to support formal methods appears to need a rich interconnection of data. This, in turn, leads to complex consistency constraints.

The system on which the authors are currently working is called "Mule". Its current status is described in Wills et al (34). Mule is being designed to support the rigorous development of software using formal methods for specification and design.

The large number of different formalisms in software engineering makes it desirable to identify common problems and to design generic tools. Each general tool is parameterized by a definition language which can be used to tailor it to particular applications. For some subproblems, these languages can be declarative in character.

This is similar to research into compiler-compilers and some of the solutions in that area can indeed be carried over. The most prominent is *syntax analysis*. In fact, this seems to be the only general solution used in most modern Programming Support Environments (PSE).

A second area where the quest for generality should be highly rewarding is that of context condition analysis. To use one extreme example: 60% – or 45000 lines – of an (un-named) Ada compiler are dedicated to static semantic checking. A very attractive solution to this problem can be found in Snelting (30).

Mule has not been designed to offer ready-made solutions to particular problems. Instead it is intended to provide basic infrastructure for a wide range of tools without commitment to

particular approaches. Mule is, in fact, fully parameterized only in
the area of syntax but it provides a coherent framework for adding new
concepts either by hand coding them or by way of parameterized
solutions.

The general part of Mule provides three components:
- a USI
- an underlying database
- a data manipulation language.

Although not easy to separate, these three items are the subjects of the
next three sections.

The current version of Mule is being applied to the support of VDM.
(The examples below are presented in a VDM style.) The initial
implementation runs on a PERQ workstation under the PNX operating
system; the programs have been written in Pascal.

3.2 STRUCTURE EDITORS

A text editor is a program which permits the creation and change of
a collection of data (i.e. the text). Such programs are normally
interactive. The simplest editors use little more information about the
data than that an end-of-line character plays a special role. A
screen-oriented editor uses the whole of a VDU screen as a window onto
the data and permits changes or insertions at any point in the window.
Other than the end-of-line, the data is treated as a sequence of
arbitrary characters. Thus it is possible to build up a file which
violates the syntax of a programming language and to have to correct
errors detected by the syntax analysis phase of a compiler. For *formal
languages* — such as programming languages — an alternative is possible.
A *Language-Oriented Editor* (LOE) limits the interaction in a way which
ensures that any documents which are created conform to a given syntax.

The form of syntax with which all computer users are familiar is the
concrete syntax (written, for example, in BNF) which governs the valid
sequences of characters for a language. In the development of
programming language semantics (cf. Lucas (21)), it has been useful to
introduce a notion of *abstract syntax*. Essentially, an abstract syntax
records the information content without the syntactic marks which are
required for parsing. The objects which conform to an abstract syntax
can be thought of as trees and correspond to the sort of structure
generated by a parser.

There are a number of advantages in building LOEs around an abstract
syntax. The need to present a linear (or concrete) version of the
abstract form of a text is handled by a *projection* (or unparsing)
scheme. The advantages of this approach are explained via a simple
example which uses the abstract syntax for logical expressions:

```
Lexpr =  Neg | Disj | Conj | Impl | Exists | Var | Rel

Neg :: N:Id

Disj :: OPD1:Lexpr   OPD2:Lexpr

Conj :: OPD1:Lexpr   OPD2:Lexpr

Impl :: ANT:Lexpr   CON:Lexpr

Exists :: BV:Id   BODY:Lexpr

Var :: Id

Rel not specified here
```

(The examples given in this paper are simplifications of those envisaged for the actual VDM version of Mule.)

The most obvious advantage of an abstract syntax basis for an LOE is the ability to employ different projection schemes. Thus a single abstract logical expression might be presented as:

    i<0 ∧ r=-i ∨ i≥0 ∧ r=i

or as:

    (i<0 & r=-i) | ((i¬<0) & r=i)

depending on the projection scheme. One bonus of this flexibility is the option to use different character sets depending on the power of the devices (printers or screens) at hand. A more subtle use of the same idea is to use alternative projection schemes in order to display the essential parts of a text which is too large to fit on the screen. The user may be given a choice of projection schemes for any particular sort of database object: for a Pascal program, one may envisage alternative projection schemes, which, for instance, bring into proximity the body of a procedure and a list of the names known within it. Abbreviating a document from its normal form in order to make visible together those parts which are of immediate interest is known as "holophraxis". The normal mode of use of a SE employs holophraxis most of the time, as one rarely wishes to see simultaneously all of the documentation which should properly be attached to any object.

Another important advantage of working with the underlying abstract form of a text is the help that this can give with layout. The work in Schwarzenberger and Zemanek (28) shows how the two-dimensional presentation of formulae can be greatly aided by the simple rule of inserting extra end-of-lines at positions in the text corresponding to the highest possible cut in the tree.

The foregoing discussion concerns single languages. It is clear how one can write a program like the 'ALOEGEN' of Gandalf which generates an LOE from an abstract syntax and one or more projection schemes. This possibility to create LOEs for new languages is crucial: the help that they provide with syntax can be quickly brought to bear on unfamiliar languages.

The next section of this paper is concerned with the underlying data model. Before turning to a more detailed consideration of some USI questions it is necessary to recognize that tree structures are not adequate for all purposes. A simple case in point is the logical expression language which is sketched above ('Lexpr'). There are clear advantages in making the instances of identifiers link together in a way which shows the quantifier bindings of the identifier occurrences in the 'BODY'. There are also many things to which Mule is to be applied which would not normaly be thought of as languages. One example, which is explored below, is proofs. In that example, it is very clear that some general graph-like structure is required as the internal representation. Another case is the storage of refinements and their relation to specifications.

Because of the richness of the underlying data structure, the form of LOE used in Mule is called a *Structure Editor* (SE). In Mule a Structure Editor is a user interface mechanism for inspecting and manipulating data structures with some non-trivial model; the data need not be a tree, and need not necessarily be thought of as the abstract representation of a statement in some language. As with the LOE, the user communicates with the system in terms of textual projections from the database; but the mapping from data to text may be somewhat more complex, and variable according to the user's requirements.

The effect of the concrete syntax on the textual appearance of a language is obvious. The way in which a language is entered via an SE can be profoundly affected by the choice of abstract syntax. Reverting to the 'Lexpr' example above, it would have been possible to write:

Lexpr = Var | Rel | Neg | Linfexpr | Exists

Linfexpr :: OPD1:Lexpr OPR:{AND, OR, IMPL} OPD2:Lexpr   etc.

This would give rise to a different sequence of options presented by the USI. It is up to the designer to choose the grammar which will suit users best!

A related, but more subtle, problem occurs with the question of type information in the abstract syntax. The division, for example, into logical and arithmetic expressions has the advantage that the menu (either of operators or permissable expression types) is restricted. If, however, all of the expressions can be - in the simplest case - identifiers, the typing information cannot actually be found by a context-free parser. At first sight, there is a temptation to claim that the ability of the user to provide the typing information is an advantage of SEs. When, however, it is recognised that - in practical systems - parsing must be supported, it is no longer clear that the claimed advantage can actually be realized.

The mapping from internal structure to visible text is the projection scheme. Clearly, for a tree editor, the concrete syntax of a language is unimportant and readily variable; in a structure editor, the projection scheme defines various views of the database and clothes information abstracted from the database in visible form. Such projections need not take textual form: there are, for example, editors which display the programs as structure charts (Giacalone et al (10)).

Commonly, the elements of a projection can be related more or less directly back to objects in the database, dependent on the complexity of the projection scheme. Co-altitudinous successive instances of 'for', ':=', 'to', 'do' would be associated with a particular node in a program tree, whilst the text in between these parts would each be associated with a subtree of the node. (As a counter-example, a projection algorithm which displays the results of a statistical analysis of the database could construct few direct correspondences into the database.) In the usual case, this relationship affords the user a means of indicating database objects to the system, by pointing at objects on the screen.

We have adopted a view of Structure Editor architecture which takes advantage of high-resolution screens and pointing devices. There are a number of windows, most of which display a particular function of the database selected by the user. Associated with each object on the screen, is a menu of operations which are applicable to the associated database objects. When the user points at the display object, the menu appears; selecting a menu item applies the operation to the relevant object in the database, and the screen is updated. The database and the screen are the only carriers of information while the system is waiting for the user (which is almost like saying it's modeless).

Language-oriented editors are restricted implementations of this general scenario: operations available on each subtree include insertions, deletions, substitutions, of subtrees of the correct types. An example of some typical features is given in Appendix A.

Whilst it can be a great help for beginners (cf. Garlan and Miller (9)), this style of document entry can be tedious for experts. It is clearly necessary to allow some textual entry (if only for identifiers). The designers of many structure editors incorporate text editing facilities and parsers. For some display objects, the menu is augmented by a set of operations which may be invoked by directly editing part of the display. Both administrative commands and database modifications can be handled this way. Not all display objects may be directly edited, because the mapping from database to screen is not one–one, and the reverse mapping is not always obvious: recall the statistical example above.

The projection scheme determines, for each display object (as a function of the database) its visible appearance; the menu of applicable operations; whether it may be edited, and if so how the edited result shall be interpreted.

Structure editors allow their users to inspect and modify complex data structures in a direct way; if the projection schemes are reasonably well–written, the user can see just those parts of the database that are of interest at any one time. All transformations made to the database, no matter what their scale, may be invoked through the structure editor, which offers at least the guidance of showing which operations are applicable to any particular substructure, and hence how it can be modified or augmented. No fixed order of work is imposed upon the user.

The tools to which the user has access through the SE may be much more closely integrated than those in a conventional environment: the rich structure of the data allows them to communicate with each other and with the user with only small–scale requirements for parsing, and with minimal overhead in accessing and generating persistent data.

Too much reliance on menus can make using the system tedious: parsers for certain common operations (such as the replacement or insertion of algebraic expressions) must be provided. Transmission of data to other systems requires projection schemes which can dump the database unambiguously. The same problem arises if fundamental changes to the grammar are proposed. The transition from a conventional system presents special problems. Whilst ideas used in the design of tools already written for conventional environments may be taken advantage of, those tools themselves will need to be substantially modified or rewritten.

Our conception of the Structure Editor loses much of its attractiveness if one tries to adapt the interface for a typewriter (through which one can only transmit peephole views of the database), or for a VDU (on which multiple windows, which facilitate modelessness, are difficult to implement effectively). Conversely, we see that bitmapped displays and pointing devices are not used to the full merely by providing multiple windows to teletype–driven processes.

The SE should be seen as being central to the toolset: other tools are incorporated into it. The close integration requirements of PSEs are satisfied by the SE–based architecture, which provides a friendly, modeless interface for the user.

3.3 REQUIREMENTS OF DATA MODEL

This section explores in more detail the data model which is evolving in the Mule project.

Given the first abstract syntax for 'Lexpr' in the preceding section, an object which matches the grammar is:

```
mk-Conj(mk-Impl(mk-Var(El), mk-Var(E2)),

      mk-Impl(mk-Neg(mk-Var(El), mk-Var(E3))))
```

It is natural to view such objects as trees:

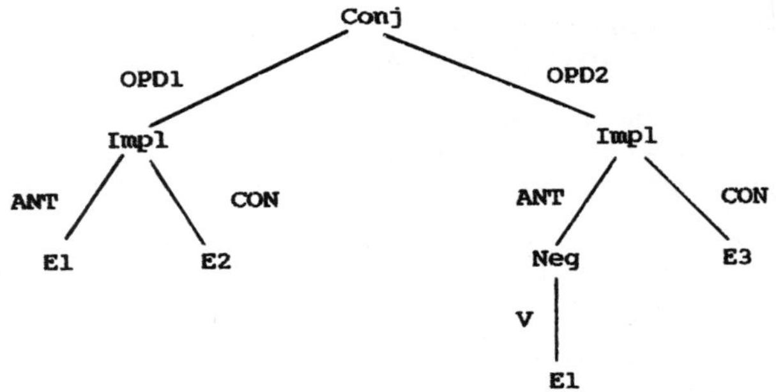

The links in this tree representation could be implemented by some form of pointer. Mule has been based on a *binary relational model* (BRM) of data. The actual database system (MDB) is described in more detail in the next section. For now, it is sufficient to understand that the basic idea of entities and binary relations has been enriched by list-like links. Instead of representing the tree by pointers, the nodes are made to correspond to entities and the branches to named relations. One consequence of this decision is that the data items are very small: MDB has to support a very fine-grained data model.

The abstract syntax itself is an object which can be stored by an SE. Here, because of the recursion on 'Lexpr', it is clear that the object is not purely a tree. If both instances and the abstract syntax are stored, they can be linked by a relation ('ISA') which shows the type of each node in the instance. Thus:

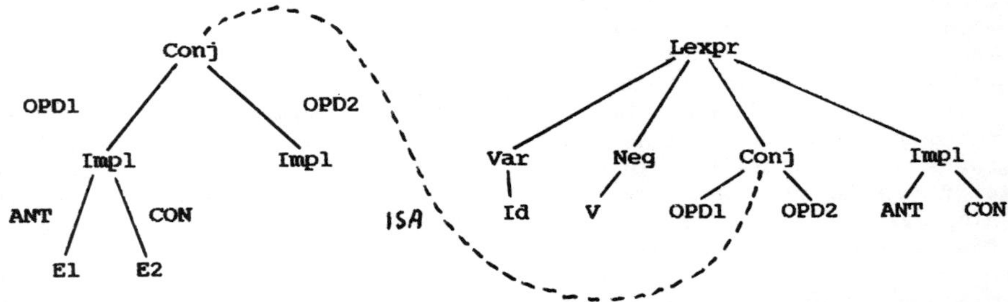

The stored AS provides the information required by the SE to display options at the USI; the instance trees are the objects which are built and changed.

The advantage of using relations for all of the links is their generality. It is possible to add new entities, or new relations to existing entities, at any time. It is also important to realize that the notion of typing (the 'ISA' relation) is not built into the underlying MDB interface. Furthermore, the graph-like quality is a natural consequence of the BRM.

Two places are identified above where a simple tree structure is inadequate: graphs are used both for the binding (in quantified

expressions) and for the AS itself. There are a number of features which Mule is expected to provide which emphasize the need for the graph data structure. It is intended to support:
- formula manipulation
- type checking
- proof construction

It should be clear from the example of bound variables in quantified expressions, that the graph-like links would greatly aid the normal substitution problem. In fact, it is argued below that, even in the simple instance of a conjunction given above, the instances of 'El' should be linked.

Type checking is required for both programming languages and for logical expressions. The task of checking that the use of variables matches the declarations in Pascal is handled in compilers by building a dictionary. In formal language definitions, an abstraction of this dictionary is a static environment. It would be possible to simulate this environment object in MDB. The approach adopted in Mule, however, is to link directly occurrences of uses of identifiers to the corresponding declarations.

The arguments above which indicate a preference for the graph-like data structure become overwhelming when the needs of proof construction are considered. The seminal systems which exist in this area (cf. Gordon et al (13), Lescanne (20), Boyer and Moore (5), Nakajima and Yuasa (25), Good (12)) are all extremely difficult to use. Almost all of these systems have been designed with a teletype-like interface. The material which must be manipulated in a proof is so extensive that the advances in large bit-mapped displays must be fully utilized. It is hoped that dramatic improvements in usability would result from improving their USIs.

The Mule work in this area is still at an early stage. The example which is given here attempts to show how a natural deduction style proof can be constructed. A possible AS for proofs is:

Theorem :: HYP: _list of_ Lexpr

       PROOF: _list of_ Step

       CON: Lexpr

Step = Theorem | Juststep

Juststep :: WHAT: Lexpr

       WHY: Rule

       FROM: _list of_ (Lexpr | Theorem)

The example proof in Appendix B shows:

$E1 \wedge E2 \vee {\sim}E1 \wedge E3 \vdash (E1{=}E2) \wedge ({\sim}E1{=}E3)$

The axiomatization used is one which accepts the notion of partial functions (cf. Barringer et al (1)). When applying a rule like "and elimination" ('∧-E') there are two choices, either the tree is copied or links are built. In order to keep control of the information, the approach to be followed in Mule would always form links.

3.4 DATA OBJECTS

One of the major influences in the adoption of the BRM by the Mule group was the NDB system (cf. Winterbottom and Sharman (35), Sharman and Winterbottom (29)). NDB was designed as an easy-to-use relational database system for modest amounts of data. A formal specification and development of NDB are recorded in Welsh (33). It was aimed at easy information retrieval and this led to a number of architectural decisions which were not necessarily appropriate for Mule. For example, all relations are stored as pointers in both directions in order to avoid inconsistencies in performance. The type model included decisions such as each entity belonging to one entity set and that relations are typed via the entity sets.

The decision to design MDB was motivated by removing architectural decisions for which it was not absolutely clear that Mule had need. In some senses, MDB is a lower-level interface than NDB. It was envisaged, and has proven to be the case, that other levels of interface would be built above MDB.

The development of the data model started out from the binary relational model: the database is a collection of entities and binary relations between entities. Entities can have values; relations are themselves represented by entities, thus allowing relations between relations. This model is very similar to the entity relationship model and semantic nets. It is completely general and well suited for storing information with many interrelationships. Furthermore, it facilitates dynamic growth by permitting the creation of new entities and relations. However, it is too general for the specific purposes of the Mule system. We do not want a knowledge base with a general querying facility but — rather — a database for a particular application.

The main role of the database is to store textual information in some graph-like form. Neither queries nor updates are arbitrary but follow a certain pattern which is dictated by the input of information via the USI. In most cases, it is not necessary to store relationships in both directions but only in one.

These considerations lead to the final data model as embodied in MDB, the Mule database. It uses unidirectional rather than bidirectional connections between entities. Labelled directed graphs, where nodes and arcs are labelled by entities and nodes may also contain values, were chosen. There is a one-one correspondence between nodes and entities, and a many-one between nodes and values and arcs and entities. (Entities and nodes can be used synonymously.)

There are a few additional concepts of which the most important one is that of lists. Frequently, it is necessary to store ordered relationships between entities (e.g. a procedure and its parameter list, a sentence and its words). Although this can always be expressed by a chain of links from one entity to the next, it is certainly more convenient to have lists available explicitly. Therefore MDB also allows links from an entity to a list of entities. A formal description of MDB can be found in (34).

In order to exploit the full potential of the data model, all information should be stored so as to expoit graph (i.e. not-tree) links to the maximum. This immediately raises the question of how this is to be achieved. One way is by explicitly indicating connections using the pointing device of the USI. It should be possible to point at a variable instead of entering its name textually. However this does not answer the question for textual input. This could be solved using an

idea called "dynamic syntax" (Hanford and Jones (14)). It allows the
association of actions, with grammar productions, which can change the
grammar. When parsing, the declaration of a variable 'I', a production:
    <expr>::= I
together with an action of linking 'I' to its declaration, is added.
The next time that the parser comes across 'I' in the context of an
expression, no new entity need be created but the parser performs the
action of linking it to the declaration of 'I'. The complementary
problem arises when updating an occurrence of an identifier: should all
occurrences be changed or only this particular one?

The generality of the BRM is described in the database literature.
The convenience of the underlying model has been borne out by the Mule
work. There is also optimism that the SE interface to MDB will prove
general.

3.5 THE DATA MANIPULATION LANGUAGE

As indicated in the introduction, Mule is not intended to solve the
particular problems of a PSE — rather it provides a means of
implementing solutions in a coherent way. Any tool built on Mule will
have to access the database. The implementation of MDB is in Pascal and
consists of a set of procedures for traversing and manipulating graphs.
These procedures are very low level: access is of the entity-at-a-time
kind. Worst of all, there is no connection to the only data definition
language in Mule, the abstract syntax. MDB presents the programmer with
an amorphous collection of nodes and arcs. Clearly, this is not an
ideal environment for implementing a complicated task like type
checking. A high-level language called GRAPL has, therefore, been
designed for manipulating MDB objects.

The main features of GRAPL were dictated by the applications under
consideration. It should support both query and update; it has to deal
with an evolving database whose structure may change over time; most
applications are of the symbol manipulating kind (e.g. type checking,
theorem proving, program transformation). Thus one requirement is for
pattern matching using the constructors from the abstract syntax. It
should make use of declarative language features thus permitting the
implementation of general tools which can be tailored to particular
formalisms very easily.

These requirements are most easily met by a functional or logic
language with special features for updating the database. It appeared
unwise to embark on the design of a completely new language and safer to
aim for some minor modifications to an existing one. The obvious
candidate is Prolog: it offers most of the required features, has been
used successfully in AI and database applications and is easy to
implement.

The main difference between GRAPL and Prolog is the underlying
domain of data: graphs instead of trees. This is reflected in the kind
of patterns used. Patterns in GRAPL have the following form:

    Pattern = <u>list</u> <u>of</u> Arc | Label | Value | Var

    Arc :: LABEL: Label TARGET: Pattern

Labels are identifiers and stand for particular entities. They
serve as the link between an abstract syntax and a GRAPL program
manipulating its instances. Given the example 'Lexpr', one could use
the labels 'OPD1' and 'OPD2' as the arcs connecting a sub-expression to

its two sons. Matching is performed according to the following rules: a variable matches anything; a label matches only the entity it represents; a value matches any node containing that value; a list of arcs matches any entity which sources at least all of the arcs in the list such that the targets match with the respective TARGET patterns. The last rule is especially important because it makes it possible to write programs that are not invalidated by the addition of further arcs to a graph. This is an essential property because of the need to extend the information in the database without having to change existing programs. As an example, take a type checker for some programming language — this type checker should work independently of any additional information subsequently attached to a program.

Patterns can therefore be considered as incomplete graphs, whereas the information in the database is complete (the closed world assumption).

The above definition of matching works given a restriction to acyclic graphs. There is, however, a generalization which handles cyclic graphs as well. This has the consequence that 'x' and 'f(x)' are unifiable, resulting in a circular structure. Although Prolog performs no occurrence check, this can be undesirable.

As Mule does not deal with graphs all the time, straight Prolog patterns are allowed as well. As trees are special kinds of graphs, this is merely a notational extension. A pattern 'Disj(t1, t2)' can be transformed into '<ISA: And, u1, u2>'. 'ISA' and 'And' are labels denoting the 'ISA' relation and the 'And' node in the abstract syntax; 'u1' and 'u2' are the transformed patterns of 't1' and 't2'. For an example, see (34), which presents a tautology checker for propositional calculus as a GRAPL program.

From the Mentor group, there has emerged a language called Typol for specifying context conditions. In flavour, it is also similar to Prolog. Their aims are very similar to ours, only restricted to the area of type checking.

Another, seemingly very different, language to describe context conditions is used in PSG. It is at one end of the spectrum, aiming to perform a particular task as well as possible. GRAPL is at the other end, striving for generality but less impressive on special tasks.

3.6 CURRENT RESEARCH

The support provided by a conventional operating system for a fine-grained database is poor. Most operating sytems embody a division into permanent "files" and "programs", and temporary "variables" — these only exist during the execution of a program. It is assumed that a program uses many more variables than files, and consequently a relatively large overhead for each file is considered acceptable. This, combined with the fact that files usually have little or no internal structure, makes the implementation of a store for a large number of permanent, small objects difficult.

Conventional languages reinforce this division by having transient variables, that come and go with the execution of a program, and a separate interface for files. In contrast, MDB is suited to a one-level persistent storage mechanism, in which entities in the database are as accessible as variables in a conventional programming language. In MDB entities have the additional property of persistence. Systems that integrate the language with the operating system Teitelman and Masinter (31), Goldberg and Robson (11), seem to achieve more success in this respect.

A particularly pointed example, of the way in which the division between permanent and temporary data has become entrenched, occurs in virtual memory systems.   It is usually assumed that data possesses locality of reference (i.e. that objects close to each other in the virtual address space of a process are likely to be related).   When a program does not possess this property (as with MDB), the virtual memory system has a negative, rather than positive effect on the performance of a program.

The current implementation of MDB loads the entire database from a (small) set of files into the user's virtual address space at the start of the session and writes it back out at the end.   Thus a fine-grained graph structure is arbitrarily mapped onto the pages of conventional virtual memory system.   When the size of the graph exceeds the size of the real memory, thrashing usually results.   To avoid this happening, and to avoid the large start-up delay inherent in such a scheme, a different form of virtual memory is required.   Solutions to the paging problem exist Kaehler and Krasner (19), Ungar (32), but they frequently make the provided paging hardware redundant.   In short, many architectural trends of the last two decades are in opposition to the requirements of a system such as MDB.

In MDB it is easy for the user to alter the arcs of the graph in such a way that entities in the database become inaccessible.   At present this "garbage" is only collected by an "off-line" mark-sweep collector (which is invoked when the graph is transferred from the user's virtual address space to a set of files at the end of the session).   However, there exists a primitive MDB function that allows the user to delete existing entities from the database; it is assumed that the user will not create "dangling" references.   For large databases the use of an off-line mark-sweep system (off-line in the sense that the database is unusable during garbage collection) is not a practical proposition.   Futhermore, it is unwise to rely on the correctness of user programs to ensure internal consistency, and unfair to place the storage management task on the user.   To provide automatic garbage collection, we envisage a scheme in which entities are reference counted to determine when they become garbage, with a background mark-sweep collector collecting circular garbage.

Current work includes extending Mule to be multi-access.   At the lowest level, this requires the introduction of locking primitives which ensure that a user has exclusive access to an entire sub-graph.   This will enable higher levels to ensure consistency and integrity of related structures.

It is hoped that the final system will be flexible enough to enable parts of the database to reside on separate machines.   The potential benefits of this arrangement include the following:
 - information can be held in the places where it is used most
 - replication of data can aid in backup and crash recovery
 - performances can be maintained as the size of the database increases

ACKNOWLEDGEMENTS

The influences on Mule from the work on personal workstations and systems like Gandalf, Mentor and PSG is obvious.   The authors are grateful to the organisers of the York IPSE conference for the invitation to present this work.   Thanks are due to Julie Hibbs for typing the paper.

REFERENCES

1. Barringer, H., Cheng, J.H., and Jones C.B., 1984, 'A Logic Covering Undefinedness in Program Proofs', <u>Acta Informatica</u> <u>Vol 21</u> <u>Part 3</u>, 251-269.

2. Bauer, F.L., and Woesner, H., 1983, 'Algorithmic Language and Program Development', Texts and Monographs in Computer Science, Springer-Verlag.

3. Bjorner, D. and Oest, O.N., (eds.), 1980, 'Towards a Formal Description of Ada, <u>LNCS</u> <u>98</u>, Springer-Verlag.
    {This is the published version which was revised during the development of the DDC Ada compiler (to match the language changes); the DDC are now producing a new formal definition of Ada under an EEC contract.}

4. Bjorner, D., and Jones, C.B., 1982, 'Formal Specification and Software Development', Prentice-Hall International.

5. Boyer, R.S. and Moore, J.S., 1979, 'A Computational Logic', Academic Press, New York.

6. Donzeau-Gouge, V., Kahn, G., Lang, B. and Melese, B., September 1983, 'Outline for a Tool for Document Manipulation' IFIP Conference, Paris.

7. Ehrig, H. and Mahr, B., 1985, 'Fundamentals of Algebraic Specification, Equations and Initial Semantics', <u>EATCS Monographs on Theoretical Computer Science</u>, <u>6</u>, Springer-Verlag.

8. Floyd, R., 1979, 'The Paradigms of Programming', <u>CACM</u> <u>8</u> 22.

9. Garlan, D.B. and Miller, P.L., May 1984, 'GNOME: An Introductory Programming Environment Based on a Family of Structure Editors', <u>ACM Softw. Eng. Notes</u>, <u>9(3)</u>, 65.

10. Giacalone, A., Rinard, M.C., Doppner, T., May 1984, 'IDEOSY: An Ideographic & Infrastructure Program Description System', <u>ACM Softw. Eng. Notes</u>, <u>9(3)</u>, 15.

11. Goldberg, A. and Robson, D., 1983, 'Smalltalk-80: The Language and its Implementation', Addison-Wesley.

12. Good, D.I., 1982, 'The Proof of a Distributed System in Gypsy', University of Texas Technical Report 30, Austin, Texas.

13. Gordan, M., Milner, R. and Wadsworth, C., 1979, 'Edinburgh LCF', <u>LNCS</u> <u>78</u>, Springer-Verlag.

14. Hanford, K.V. and Jones, C.B., 1974, 'Dynamic Syntax: A Concept for the Definition of the Syntax of Programming Languages' pp115-142 in 'International Tracts in Computer Science and Technology and Their Application', Permagon Press.

15. Hayes, I.J., February 1985, 'Applying Formal Specification to Software Development in Industry', <u>IEEE Trans. on SW. Eng.</u>, <u>11(2)</u>, 169-178.

16. Jones, C.B., 1980, 'Software Development: A Rigorous Approach', Prentice-Hall International.

17. Jones, C.B., 1983, 'Systematic Program Development' Invited paper at Symposium 'Wiskunde en Informatica', CWI Amsterdam, Holland.

18. Jones, C.B., 1984, 'Specification, Verification and Testing in Software Development', invited paper at Workshop for 'Software: Requirements, Specification and Testing', April, Norwich, England.

19. Kaehler, T. and Krasner, G., 1983, 'LOOM — Large Object-Oriented Memory for Smalltalk-80 Systems', pp251-272 in 'Smalltalk-80: Bits of History, Words of Advice' G. Krasner (Ed.), Addison-Wesley.

20. Lescanne, P., 'Computer Experiments with the REVE Rewriting System Generator' <u>POPL '83</u>, Austin, Texas.

21. Lucas, P., 1982, 'Main Approaches to Formal Specifications', Ch. 1 in (4), pp3-24.

22. Medina-Mora, R., Notkin, D.S., November, 1981, 'ALOE Users' and Implementors' Guide', Carneige-Mellon University Report, CMU-CS-81-145, Pittsburgh, Pittsburgh PA, USA.

23. Meyer, M. and Nerson, J.-M., 1984, 'A Visual and Structural Editor', University of California, Santa Barbara, TRCS84-03.

24. Morgan, C.C. and Sufrin, B.A., March 1984, 'Specification of the Unix Filing System', <u>IEEE Trans. Software Eng.</u>, <u>SE-10</u>, pp128-142.

25. Nakajima, R., and Yuasa, T. (Eds.), 1983, 'The IOTA Programming System, A Modular Programming Environment' <u>LNCS</u>, <u>160</u>, Springer-Verlag.

26. Partsch, H., April 1984, 'Structuring Transformational Developments: A Case Study Based on Earley's Recognizer', <u>Science of Comp. Programming</u>, <u>4(1)</u>.

27. Reif, J.H. and Scherlis, W.L., December 1982, 'Deriving Efficient Graph Algorithms', Carnegie-Mellon Technical Report no. CMU-CS-82-155.

28. Schwarzenberger, F. and Zemanek, H., 28 July 1966, 'Editing Algorithms for Texts over Formal Grammars', IBM Laboratory Vienna, Technical Report TR25.066.

29. Sharman, G.C.H. and Winterbottom, N., 1978, 'The Data Dictionary Facilities of NDB', in <u>Proceedings of the 4th VLDB</u> 186-197.

30. Snelting, G., 1985, 'Experiences with the PSG-Programming System Generator', TAPSOFT Joint Conference on Theory and Practice of Software Development, Berlin.

31. Teitelman, W., and Masinter, L., April 1981, 'The Interlisp Programming Environment', IEEE COMPUTER, 14(4), 25–33

32. Ungar, D., April 1984, 'Generation Scavenging: A Non-disruptive, High-performance Storage Reclamation Algorithm', ACM SIGPLAN Notices, 19(5), 157–167, Pittsburgh PA.

33. Welsh, A., 1982, 'The Specification, Design and Implementation of NDB', M.Sc. Thesis, Dept. Computer Science, Manchester University, England.

34. Wills, A.C., Wolczko, M., Nipkow, T., Jones, C.B. and Cottam, I.D., 1984, 'Mule – An Environment for Rigorous Software Development', Second Annual Report, Department of Computer Science, Manchester University, England.

35. Winterbottom, N. and Sharman, G.C.H, September 1979, 'NDB:Non-programmer Data Base Facility', IBM UK Technical Report TR.12.179.

APPENDIX A: ILLUSTRATION OF A STRUCTURE EDITOR

This Appendix illustrates features which are needed in a structure editor. (Readers who are familiar with SDEs are advised to study the more interesting example in Appendix B.) The example, here, involves the creation of a Pascal fragment. Assume that – at some point – the window shows:

    program thing (input, output);

     begin [Stmt] end.

    ----

     begin if case while repeat for with := call

The text contained in the special brackets ('[...]') is the name of a syntactic class – it marks a "hole" whose contents the user can generate; the underlining represents some kind of highlighting – the text has a nested block structure, and the user can move the (highlighted) cursor onto any block; the local menu indicates the possible options for expansion of the current syntactic class (there is a global menu in a separate window). The user points at '_repeat_':

    program thing (input, output);

     begin

      repeat [Stmt] until [Condition];

      [Stmt]

      end.

     ----

     begin if case while repeat for with := call

Note that the cursor has been repositioned to where the user is likely to want to work. The user can type text, which will appear at the immediate left of the hole, and will be parsed when he hits 'RETURN'. The parser will accept incomplete expressions, and expressions containing the character '⊥', which denotes an undeveloped construct. Syntax errors leave the text unchanged so that the user can correct. The user types: 'if x<y then ⊥ else r:=x [RETURN]'

    program thing (input, output);

     begin

      repeat

       if x<y then [Stmt] else r := x

      until [Condition];

      [Stmt]

      end.

     ----

     begin if case while repeat for with := call

User types 'begin [RETURN]'. The parser accepts unambiguouous partial

```
input of this sort.

    program thing (input, output);
     begin
      repeat
       if x<y then
          begin [Stmt] end
        else r := x
      until [Condition];
      [Stmt]
     end.
    _________________________________________

     begin if case while repeat for with := call
```

Note pretty printer rearranges over-length lines. The User types
'r := y[RETURN]':

```
    program thing (input, output);
     begin
      repeat
        if x<y then
           begin
            r := y;
            [Stmt]
           end
         else r := x
      until [Condition];
      [Stmt]
     end.
    _________________________________________

     begin if case while repeat for with := call
     delete
```

The body of a compound statement can be a list of statements — for such
lists, a continuation item is generated automatically; the 'delete' menu
item is provided to terminate such lists. The user types '[RETURN]',
which is equivalent of selecting 'delete', and does not invoke the
parser.

```
    program thing (input, output);
     begin
      repeat
        if x<y then
           begin
            r := y
           end
         else r := x
      until [Condition];
      [Stmt]
     end.
```

```
    not and or = <= >= < > in factor
```

The user moves the cursor to 'program'. The new menu includes items for inserting the currently empty fields.

```
    program thing (input, output);
     begin
       repeat
        if x<y then
           begin
            r := y;
           end
         else r := x
      until [Condition];
      [Stmt]
     end.
```

```
    const type var module
```

Select 'var'

```
    program thing (input, output);
     var [Id] : [Type];
     begin ... end.
```

Here, the body of the procedure has been replaced by ellipsis points – this example of "holophraxis" is not necessarily realistic and is shown only as a reminder of the feature.

APPENDIX B: HANDLING PROOFS

This Appendix shows how the Mule Structure Editor might be used in proof
generation.   Suppose – at some stage – the screen has the following
projection of a trivial specification:

<u>operation</u>

[Name] ([Op-signature])

<u>ext</u> [Name]

<u>pre</u> [Predicate]

<u>post</u> [Predicate]

<u>decomposition</u> [Decomposition]

———————————————————————————

(Operation name)

Interactions – like those shown in Appendix A – might create:

<u>operation</u>

ABS

<u>ext</u> <u>rd</u> I: Real <u>wr</u> R: Real

<u>pre</u> <u>true</u>

<u>post</u> (i<0 ⊣ r=–i) ∧ (¬(i<0) ⊣ r=–i)

<u>decomposition</u>

D1: <u>begin</u>

    <u>pre</u> <u>true</u>

    <u>if</u> i<0 <u>then</u> r := –i <u>else</u> r := i

    <u>post</u> (i<0 ⊣ r=–i) ∧ (¬(i<0) ⊣ r=–i)

    <u>end</u>

———————————————————————————————————

Pascal-vcg ...

The pre– and post–conditions would be generated automatically from the
specification.   The graph connections in the database are being used
heavily here: expressions like 'i<0' only occur once; the projection
scheme displays the expressions in full in each place they are used.
'Pascal-vcg', is a sophisticated operation which uses the Pascal
statement and generates a proof obligation, which might be:

    <u>from</u> i<0 ∧ r=–i ∨ ¬(i<0) ∧ r=i

    <u>infer</u> (i<0 ⊣ r=–i) ∧ (¬(i<0) ⊣ r=–i) [<u>Justification</u>]

——————————————————————————————————————

cases ∧-introduction

The prompts indicate that a proof could be undertaken. Although
relatively straightforward, the result required here could be used as a
derived rule of the logic. The next stage might be the creation of a
general form such as:

```
    from  E1∧E2 v ~E1∧E3

    ·infer (E1⇒E2) ∧ (~E1⇒E3)                    [Justification]
    -----------------------------------------------------------------
```

cases ∧-introduction

The prompts indicate proof rules which are applicable – the ones used
here all relate to the Logic of Partial Functions in Barringer(1); later
stages of this proof also use some obvious derived rules. The user
chooses 'cases':

```
    from  E1∧E2 v ~E1∧E3
1            from  E1∧E2

             infer (E1⇒E2) ∧ (~E1⇒E3)           [Justification]


2            from  ~E1 ∧ E3

             infer (E1⇒E2) ∧ (~E1⇒E3)           [Justification]
      infer (E1⇒E2) ∧ (~E1⇒E3)                  v-E(h, 1, 2)
    -----------------------------------------------------------------
```

∧-elimination ∧-introduction

The word 'cases' is a mnemonic for the 'v' elimination rule. It is
applicable because the main hypothesis is a disjunction and it has the
job of implementing a 'forward' proof using the 'v-elimination' rule.
It breaks up the hypothesis and "copies" it with the conclusion (in
fact, the two inner hypotheses are just references to the subtrees of
the main hypothesis and the copies of the conclusion are pointers). The
task has shifted to the justification of two inner proofs. The user now
works forward choosing '∧-elimination' and pointing to line 1:

```
    from  E1∧E2 v ~E1∧E3
1            from  E1∧E2
1.1              E1                              ∧-E(h1)
1.2              E2                              ∧-E(h1)
             infer (E1⇒E2) ∧ (~E1⇒E3)           [Justification]


2            from  ~E1 ∧ E3

             infer (E1⇒E2) ∧ (~E1⇒E3)           [Justification]
      infer (E1⇒E2) ∧ (~E1⇒E3)                  v-E(h, 1, 2)
    -----------------------------------------------------------------
```

∧-introduction ...

The user can probably see how to create the necessary conjuncts but, in order to illustrate how the USI can be used to generate further sub-goals, assume that the user elects to justify the conclusion of 1 by the introduction of '∧'. Two new sub-goals are generated:

<u>from</u> E1∧E2 ∨ ˜E1∧E3

1	<u>from</u> E1∧E2		
1.1		E1	∧-E(h1)
1.2		E2	∧-E(h1)
1.3		E1 ⇒ E2	⟦Justification⟧
1.4		˜E1 ⇒ E3	⟦Justification⟧
	<u>infer</u> (E1⇒E2) ∧ (˜E1⇒E3)		∧-I(1.3, 1.4)

2	<u>from</u> ˜E1 ∧ E3	
	<u>infer</u> (E1⇒E2) ∧ (˜E1⇒E3)	⟦Justification⟧
<u>infer</u> (E1⇒E2) ∧ (˜E1⇒E3)		∨-E(h, 1, 2)

   ...

Line 1.3 is justified by a rule which introduces a vacuous implication:

<u>from</u> E1∧E2 ∨ ˜E1∧E3

1	<u>from</u> E1∧E2		
1.1		E1	∧-E(h1)
1.2		E2	∧-E(h1)
1.3		E1 ⇒ E2	⇒-vacl(1.2)
1.4		˜E1 ⇒ E3	⟦Justification⟧
	<u>infer</u> (E1⇒E2) ∧ (˜E1⇒E3)		∧-I(1.3, 1.4)

2	<u>from</u> ˜E1 ∧ E3	
	<u>infer</u> (E1⇒E2) ∧ (˜E1⇒E3)	⟦Justification⟧
<u>infer</u> (E1⇒E2) ∧ (˜E1⇒E3)		∨-E(h, 1, 2)

   ...

The next step is similar but requires that a negated form of the antecedant of 1.4 is established – this is achieved by '˜-I'; this causes some renumbering which is trivial to implement (everything internally is located by pointers – the numbering is part of the projection scheme.

```
        from  E1∧E2 ∨ ~E1∧E3
1               from  E1∧E2
1.1                   E1                              ∧-E(h1)
1.2                   E2                              ∧-E(h1)
1.3                   E1 ⇒ E2                         ⇒-vac1(1.2)
1.4                   ~~E1                            ~~-I(1.1)
1.5                   ~E1 ⇒ E3                        [Justification]
                infer (E1⇒E2) ∧ (~E1⇒E3)             ∧-I(1.3, 1.5)


2               from  ~E1 ∧ E3
                infer (E1⇒E2) ∧ (~E1⇒E3)             [Justification]
        infer (E1⇒E2) ∧ (~E1⇒E3)                     ∨-E(h, 1, 2)
```

The final proof would be:

```
        from  E1∧E2 ∨ ~E1∧E3
1               from  E1∧E2
1.1                   E1                              ∧-E(h1)
1.2                   E2                              ∧-E(h1)
1.3                   E1 ⇒ E2                         ⇒-vac1(1.2)
1.4                   ~~E1                            ~~-I(1.1)
1.5                   ~E1 ⇒ E3                        ~⇒-vac2(1.4)
                infer (E1⇒E2) ∧ (~E1⇒E3)             ∧-I(1.3, 1.5)


2               from  ~E1 ∧ E3
2.1                   ~E1                             ∧-E(h2)
2.2                   E3                              ∧-E(h2)
2.3                   E1 ⇒ E2                         ~⇒-vac2(2.1)
2.4                   ~E1 ⇒ E3                        ⇒-vac1(2.2)
                infer (E1⇒E2) ∧ (~E1⇒E3)             ∧-I(2.3, 2.4)
        infer (E1⇒E2) ∧ (~E1⇒E3)                     ∨-E(h, 1, 2)
```

The program development could be resumed by instantiating this new
derived rule.   (It is fun to construct a diagram of some of the
database!) Notice that ordinary editing operations are not available
for the steps of a proof. You can change a name (automatically changing
all the places in which it appears), and dismantle a proof in an order
which leaves no inconsistencies, but you can not just change the lines
arbitrarily. (What you can do is build a new version starting with the
existing one: but that's a separate − version control − story.) The
editing operations and rules illustrated here can easily be programmed
in Pascal or GRAPL.   It is less obvious either how to incorporate
derived rules or how to organise menus for vastly augmented rule sets.

Chapter 4

A facts and rules based approach to generalised attribute management

O.P. Brereton and P. Singleton

4 ABSTRACT

This paper describes the application of formal techniques of knowledge representation and manipulation to the management of attributes in a UNIX environment. A uniform programming model is described for both the lower-level attributes and characteristics of the filestore itself, and for some of the higher-level functionality that the shell adds. Many informal user conventions are elegantly superceded by this model, and examples are given, both for the re-implementation of existing ad-hoc features, and of new features that may be added.

4.1 INTRODUCTION

This work forms part of a wider project (EXPRESS) aimed at the application of IKBS techniques to software engineering. EXPRESS aims to capture and represent a significant proportion of the knowledge and skills that an expert software engineer uses during the various stages of the software lifecycle, so that machine-based tools may be constructed with specialised problem-solving expertise.

4.1.1 The EXPRESS project

A major objective for Software Engineering over the next few years is the establishment of "Information System Factories" in which automation and large scale capital investment will replace today's craft-based techniques and enable the faster and cheaper production of reliable software. In the EXPRESS project at the University of Keele, we are undertaking research into the application of knowledge-based techniques to automate defined tasks within the software engineering lifecycle. The main emphasis of the work is directed at better support for the maintenance of computer software – an often neglected area, but one which has been reported to consume the majority of the overall software lifecycle budget.

The fundamental problem being addressed is the evolution of a software product. At present, the modification and development of a software product is often difficult, and sometimes impossible. Changes can have uncontrolled effects unless all dependencies and interrelationships are explicitly recorded in such a way that the consequences of a modification can be fully defined or inferred.

Although the adoption of formal methods, particularly in the specification stages of the lifecycle, can be expected to improve product reliability, software will still change with time for the following reasons:

a) the specification will evolve, driven by changing customer and marketplace needs.

b) the host environment in which the software is developed will change.

c) the demands by and support for project management tools will change.

We hope to show that the use of IKBS techniques offers three principal advantages over "non-intelligent" tools.

i) <u>flexibility</u>: we hope to show that tools can be tailored to meet the needs of different users (e.g. a large DP department or a small systems project).

ii) <u>reduced costs</u>: software evolution can be managed in a controlled way, without unexpected problems. Routine supporting clerical work may also be automatically handled by the toolset.

iii) <u>reduced timescales</u>: it is vital to maximise the re-use of existing software (which itself is subject to evolution).

4.1.2 Software engineering and knowledge-based techniques

Despite well-documented successes in a range of application areas, expert systems have made little impact on the software engineering world. Nevertheless, much software is produced out of budget and timescale, with many released errors and excessive maintenance costs. This imposes an urgent demand for better tools and methods. There is a welcome move towards the adoption of more formal methodologies. It is, however, the success of expert systems in areas such as planning, control, design and diagnosis that has stimulated our interest in their potential application to software engineering, and in particular to configuration management.

We argue for a knowledge based approach for three principal reasons:

i) Traditionally, a considerable volume of knowledge about the state of a developing system is committed only to human memory, or denoted by ad-hoc conventions whose semantics are committed only to human memory. Our approach encourages and exploits the explicit representation of both formal and informal knowledge in a fact/rule base. As a very simple example, if the dependency between a document X and program module Y is not stored explicitly, their consistency cannot be maintained automatically, nor can their inconsistency even be detected automatically.

ii) the knowledge base and the inference engines are held separately: one attraction of a rule-based approach is that the problem-solving techniques common to the various applications can be separated from the rules which characterise that application. Thus adaptation

involves altering only a nucleus of relevant material, and it is
not necessary to understand or alter the problem-solving and inference algorithms themselves.

iii) the available knowledge is likely to be incomplete, and non-
deterministic techniques are needed to cope with the uncertainty.

We feel that a particularly important problem is to formalise and
homogenise the communication of knowledge between the developers and the
maintainers of a system. In most cases, the designers and specifiers of
a system will not be the same personnel who must later maintain the product. Over the lifetime of a piece of software, the maintenance personnel will also change. Thus knowledge which can be captured and
represented can potentially reduce maintenance costs, and avoid the
situation in which software cannot be altered, for fear of "bringing
down the house of cards".

4.1.3   Overview of paper

The work described in this paper represents the first part of a two
stage feasibility study. The overall objective of the study is to inves-
tigate how knowledge-based techniques may be used to represent and process knowledge that currently is implicit or stored only in human
memory. We have chosen the UNIX+[1] environment for the study, and Edinburgh Prolog [2] as the implementation language. The more detailed
objectives of the research are described in section 2.

The first part of the feasibility study has taken as its boundary
the facilities that already explicitly exist in a UNIX Version 7 system. The approach has been to reimplement a number of UNIX facilities,
such as filestore naming and permissions, using Prolog. This has enabled
us to compare the two approaches and to draw conclusions about the
representation of a relatively simple and well-understood knowledge
base. It has also provided valuable insight to the use of Prolog in this
application area, in particular showing the limitations of encapsulating
a Prolog-based tool in a UNIX environment. Section 3 of this paper
describes the implementation undertaken and investigations made, while
section 4 describes further potential applications that have been identified in the course of the study. In section 5 an assessment of Prolog
in this application area is described, and conclusions are drawn.

4.2   OBJECTIVES

In Prolog, knowledge is represented using a database of facts and
rules. A fact is a predicate with one or more arguments, representing
some external relationship. A rule is a general statement about objects
and their relationships. It is a clause in the database with functor
":-", read as "if", and two arguments, the rule head and the rule body.
Having set up a knowledge base, we can provide a goal which Prolog
attempts to prove using inference. A feature of the scheme is that the
inferencing engine (which is problem independent) is separate from the
knowledge base (which is problem specific). Knowledge representation is
an intensively studied field, and it is not the intention to review it

---

+UNIX is a trademark of Bell Laboratories

here. Rather, we shall identify in more detail the objectives of the feasibility study of which the work described in this paper forms part.

4.2.1 Applicability

In UNIX, "knowledge" is already encoded in system data structures and programs. We are interested to determine what UNIX facilities could be usefully represented in terms of a Prolog knowledge base, and then to compare the two. We wish to assess the generality of Prolog as a pro-gramming language, and determine any serious impediments to its use in 'real' applications.

4.2.2 Packaging and interfaces

There is a general problem of packaging inference. Should it be as a program, a library function, or a kernel facility? How intimately can it be interfaced to other programming languages? There will surely always be tasks which other languages could perform better than Prolog, and even if there were not, those languages would still be used. Furth-ermore, a mixed language programming environment imposes demands on type handling and checking, which should be preserved by an encapsulated knowledge-based tool.

4.2.3 Modularity

It is unrealistic to require that all a system's knowledge be held in a single database. Knowledge will inevitably be partitioned and scattered: for historical reasons; for physical reasons; according to application boundaries; and perhaps for efficiency of inference on knowledge subsets. Indeed, many of the advantages of modularity of software must be expected to apply to knowledge-based tools. We expect that knowledge bases will be combined (perhaps to form "virtual" knowledge bases) in several ways: by simple concatenation; by nesting; by masking; and by re-ordering. The UNIX filestore is a large enough application to focus attention on these issues.

4.2.4 Formalising informal mechanisms

Within a typical UNIX filestore, there are numerous untidy, unreli-able and ad-hoc applications of what we would term rules and facts; knowledge is imperfectly and incompletely represented, and only the sim-plest inference is performed.

The preponderance of suffices on user and system filenames shows a desire to represent facts about files to guide their manipulation. The rules that exploit these attributes are typically crude, inextensible and often unwritten. In contrast, Prolog's notion of a rule allows gen-eric relationships to be defined, that hold not only for all current knowledge, but also for knowledge added in the future. Hence a rule such as:

```
can-read(User, File) :-
        supervises(User, Student), can-read(Student, File).
```

is a very powerful mechanism for abstraction. We wish to study the applicability of this type of rule.

4.2.5 Efficiency

Prolog facts are necessarily more explicit than the flags and fields that they replace, so a greater space requirement is to be expected. On the other hand, rules can make general statements that might otherwise occupy far more space in a traditional data structure. We hope to get at least a feel for the relative efficiency of Prolog and traditional programs, and hope that optimal Prolog "code" will not run absurdly slower than its traditional equivalent (where it has one).

4.3 INITIAL INVESTIGATIONS

This section considers the use of attributes in UNIX and identifies some applications of Prolog to their management.

4.3.1 Attributes in UNIX

The term "attribute" is used throughout this document to denote simple facts about particular objects. Attributes to be found in UNIX can be categorised as follows:

       per-object extrinsic attributes, denoted within each inode
       e.g.   "read", "write" and "execute" permissions
             type of file (directory, file or device)
             times of latest modification and reference

       the hierarchical namespace (the tree of directories)

       conventional informal attributes
       e.g.   filename suffices (difficult to have more than one each)
             "magic numbers" within files

       attributes as yet unrepresented
       e.g.   "proven"
             "valuable"

An attribute is not part of the object to which it applies: it is a fact, or more appropriately an "opinion", about it. Different factbases may contain different versions of the truth.

4.3.2 Intrinsic and Extrinsic Attributes

Essentially, this is a distinction of representation. UNIX files, when used as directories, are extrinsically labelled as such. No such explicit distinction is made, however, between various flavours of user files, such as "source", "binary" or "data". Intrinsically, nevertheless, these attributes exist, although perhaps only in the minds of the users, who may not even agree ("source" for the goose may not be "source" for the gander).

The UNIX utility 'file' applies a modest expertise about both intrinsic and extrinsic attributes, and attempts to make a few useful, general observations about any specified file. The computation involved in inferring its extrinsic attributes is trivial: bit fields within the "inode" (a sort of pro forma luggage tag that almost every persistent UNIX object carries) assert, for example, that a file is executable. Intrinsic attributes are encoded less accessibly within the data itself, and 'file' may make rudimentary tests of vocabulary or character usage in an attempt to recognise (say) FORTRAN and C sources.

Where extrinsic attributes are intended to reflect intrinsic ones, they may of course do so incorrectly: not every ".c" file will compile, or even conform to C's context-free syntax. Such semantic discrepancies are not the concern of the inference mechanism.

Intrinsic attributes could conceivably be exploited by inference, since a rule might be defined which perused the contents of an object, to determine (for example) whether it is a text file. The use of "magic numbers" near the beginnings of files is an attribute representation that falls part way between the two forms.

4.3.3 Rules in UNIX

These are to be found within the kernel, within utilities, within a user's expertise, and occasionally within data files. An example from the 'make' utility [3] follows, as it appears in the documentation:

> "If two paths connect a pair of suffices, the longer one is
> used only if the intermediate file exists, or is named in
> the description."

4.3.4 Manipulating the UNIX filestore namespace in Prolog

Every file in a UNIX filesystem has a unique identifying number. Users and processes are rarely aware of these, as the hierarchical naming scheme introduces "directories" – sets of (number,name) pairs – with sufficient constraints that the pool of files is organised as a tree. Relative to the root directory of the tree, a file is identified by a sequence of names: its "pathname". Indeed, since each directory knows its parent (as ".."), every file has a pathname from every directory (up the tree to a common ancestor, then down another way). Two features complicate this model: directories know themselves (as "."),introducing (trivial) cycles; and a file (but not a directory) may be the child of more than one directory.

A factual representation of this structure is readily obtained by taking each directory entry, and making its implicit content explicit. Thus an example directory, number 1971, visualised as

```
1971 .
2993 ..
1929 c.c
1943 contains
1941 contains.sh
1942 cprolog.ifacts
1966 ifacts
1964 ifacts.c
1940 makefile
1945 usr.filecount
```

becomes

```
x(1971,contains,1971,'.').
x(1971,contains,2993,'..').
x(1971,contains,1929,'c.c').
x(1971,contains,1943,'contains').
x(1971,contains,1941,'contains.sh').
x(1971,contains,1942,'cprolog.ifacts').
x(1971,contains,1966,'ifacts').
x(1971,contains,1964,'ifacts.c').
x(1971,contains,1940,'makefile').
x(1971,contains,1945,'usr.filecount').
```

There is some perversity here which needs explanation. Standard Edinburgh Prolog notation puts the initial member of each factual tuple outside the parentheses, calls it the "predicate", and may restrict operations upon it (such as instantiating a variable in this position). This
shows the logical rather than pattern-matching origins of the language.
Because this is irregular, and goes against the ideal of making facts
accessible to the widest possible range of inference, we routinely
employ a "dummy predicate" (always "x"). The actual predicate (an atom
introduced to distinguish the facts of this relationship from those of
others in the factbase) is stored as an argument, and not necessarily
the first.

These facts are sufficiently explicit to stand alongside those for
other directories. Once an entire filesystem is converted, rules can be
employed to mimic namespace operations.

4.3.4.1 Pathname evaluation. Each process has two "handholds" on the
naming tree – its "root" directory and its "working" directory – and it
identifies files to the kernel (e.g. when opening them) by pathnames
relative to one or other of these contexts. The syntactic convention
used by the kernel is irrelevant to this exercise. The Prolog implementation is explicit:

```
path(A,B,C).
```

instantiates C to the inode number of the file (or directory) specified
by path B relative to the directory whose inode number is A. Argument B
is a list of names, and the two rules required are

```
path(A,[],A).

path(A,[B|C],E) :- x(A,contains,D,B),path(D,C,E).
```

The second rule is recursive, using the head (B) of the pathname list to
advance one step from the base directory A to the next node D as given
by the matching "contains" fact. The first rule terminates the recursion by handling the null pathname case.

This code is considerably shorter than the equivalent UNIX source
code, reflecting not only the degree to which general problem solving
principles have been abstracted into the Prolog inference mechanism, but
also the extent to which error handling has been omitted from the Prolog
version.

4.3.4.2 Directory listing. A Prolog implementation of the UNIX utility "ls" would initially use the 'path' rule to convert pathname arguments into explicit inode numbers. Beneath such argument handling, not necessarily best implemented in Prolog, there is a "raw" directory listing rule as follows:

```
ls(X) :- x(X,contains,_,Y),write(Y),nl,fail.
```

which could be articulated thus:

"write all the Ys such that X knows something by the name Y"

Two improvements are needed, before the performance of the Prolog version can fairly be compared with the venerable UNIX utility: the special entries "." and ".." that are a permanent feature of every directory, must be excluded from the output. The rules

```
exclude('.').
exclude('..').

ls(X) :- x(X,contains,_,Y),not(exclude(Y)),write(Y),nl,fail.
```

achieve this. Finally, the output of 'ls' must be in lexicographic order. Because UNIX files do not support insertion of data, only overwriting (or appending), a fixed-format directory structure is employed, with vacant slots (resulting from deletions) being used when available for new entries. Thus the entries are not maintained in lexicographic order, and any utility which wishes to make them appear so must sort them. Prolog's factbase allows insertion of facts at any specified point, and it is appropriate to maintain at least a partial ordering of 'contains' facts such that the children of any directory are stored in lexicographic order. This is not a trick, but a rational exploitation of the better functionality of the Prolog factbase over the UNIX file. The cost of inserting each new directory fact in an appropriate place is diluted by the ratio between directory reads and alterations: at least the traditional ten-to-one.

A comparison of the two implementations was made as follows: The directories of a PERQ workstation were converted to (dummy predicate) factual form, giving 2200 facts. The time and space consumptions reported by the Prolog interpreter are presented, along with the times to execute the UNIX utility and the space taken by the UNIX directories.

	UNIX:	Prolog:
bytes/entry:	19	73
cputime/ls:	0.8s	0.75s
realtime/ls:	2s	2s

The figure of 19 bytes per UNIX directory entry includes a proportion for space wasted by "holes" where files have been unlinked or deleted. Prolog's figure of 73 bytes would be around 69 bytes without the dummy predicate. The UNIX cputime does not include idle time spent waiting for blocks to be tranferred from disc, only the cpu overheads of so doing. The Prolog time seems to increase better than linearly with factbase size, but this has not been checked.

4.3.4.3 <u>Filestore-wide name searching</u>. The name by which a file is known in its parent directory might be used to search for the file throughout the filestore. This task is infeasibly expensive within UNIX, but in a Prolog factbase should be much faster. A study of a filestore (of over 20000 files) showed that 45% of all pathnames were unique in their final component.

4.3.5 <u>User-defined rules for access to files</u>

If file ownership information is held in a factbase, complex access rules could be defined. A project was undertaken to develop a Prolog-based permissions strategy [4]. The main conclusions concerned the dif-ficulty of interposing the inference between the unmodified UNIX kernel and the various utilities that attempted to gain access to files. This experience is summarised below.

4.3.6 <u>Combining the functionality of the shell and Prolog</u>

The UNIX process and interprocess communication models were inade-quate to support the abstract combination of shell and Prolog func-tionality that we wanted to implement. Essentially, we needed to com-bine a persistent instance of each, with a suitably robust and synchron-ised communication between then. Four arrangements were considered.

4.3.6.1 <u>Compilation into a single process</u>. This was impracticable for two reasons; our version of Prolog does not have an embeddable pro-cedural interface, only an interactive textual interface; and process size limitations would have prevented use of any non-trivial factbase. Aside from these local practical problems, it seems inelegant to bind both the Prolog engine and its database so tightly to each utility which it serves, when both components may be shareable in principle.

4.3.6.2 <u>Two communicating processes</u>. This arrangement makes better use of the limited process address spaces, but is let down by the unstruc-tured interprocess communication, which can only be via byte streams along "pipes". The lack of synchronisation between outgoing and incom-ing data, and the lack of any structuring mechanisms for that data, sug-gest that a protocol should be employed, but the fully suspended read and write operations prevent the use of conventional protocols. A "procedure-calling" interface to the Prolog process would provide the necessary synchronisation, but this requires a kernel modification.

4.3.6.3 <u>Prolog process establishes temporary shells as needed</u>. This is closer to the procedure-calling model, except that a new shell instance is created each time. Ignoring the extravagance, there is a problem of preserving state between shell invocations. It is possible, although clumsy, to use the filestore for this. This scheme does not so much enhance shell functionality with Prolog power and generality, as vice versa, but it has been implemented and appraised.

4.3.6.4 <u>A shell establishes temporary Prolog processes as needed</u>. Although this scheme seems more natural than the former, the expense of saving and reloading state between Prolog invocations has effectively ruled it out.

4.4 FURTHER APPLICATIONS OF FACTS AND RULES

This section describes further examples of ways in which a fact-and rule-based mechanism could be used to advantage within UNIX. The list is by no means complete, but suggests the range of possible applications.

4.4.1 Extended naming functions under UNIX

Various enhancements to the basic namespace operations have been implemented within UNIX utilities. A notable example is the "search path" facility of the shell, which maps the first word of a simple-command onto an executable file by searching a sequence of directories (defined within each process' environment) until a matching filename is found. This convenience approximates, although poorly, to mechanisms one might perceive in human conversation, where the meaning of a term will be sought in a series of contexts common to the speaker and the listener. A fact/rule model provides opportunities for implementing still more intelligent search strategies, customised to individual requirements or preferences.

4.4.2 Supporting Virtual Objects

Consider the enquiry "How much space is INGRES taking up?", perhaps expressed as "diskusage ingres" to UNIX. The notional object "ingres" include its sources, binaries, libraries and manual entries scattered throughout the filestore. The best that UNIX can manage is "du /ingres", which reports the size of a given subtree, or "quot ingres", which scans the whole filestore to report the total file space in use by user "ingres". Neither case corresponds to the intention, but the "quot" case illustrates how an attribute can hold an abstraction together, no matter where the components are stored. This is a modest example of configuration management.

4.4.3 Beyond Procedural and Object-oriented Programming Syntax

A shell command, such as

    lp fred.raw

is not unlike a procedure call in an imperative language

    lp ("fred.raw")

in that the interpretation of the command or procedure name ("lp") is independent of any attributes (".raw") of the actual parameter passed (many languages have overloaded operator symbols, but their meaning is resolved at compile time). Different arguments will, nevertheless, need different treatment, and "lp" may well observe the ".raw" attribute and act accordingly. The algorithms for each case that "lp" distinguishes will be compiled together into the executable "lp" file. Only with difficulty can users tailor the behaviour of "lp" to their own requirements.

With N disjoint categories of file (an oversimplification) and M generic operations ("print", "edit", "archive" ..) there are potentially MxN algorithms. Adding a new file category involves updating all generic utilities: adding a new utility involves considering all file types.

The object-oriented model is complementary: the generic operations are "messages" sent to the object, which contains "methods" for each operation. In principle, this model is little different from the first, as the code for each operation/category case is indexed from each object rather than from each utility. In practice, objects carry references to code rather than code itself.

The two models differ in storing the rectangular array of algorithms by row or by column. A fact/rule implementation could subsume both models, allowing a command such as "lp fred.raw" to be matched against several (e.g. private and public) rulebases. To modify the interpretation of commands need not involve recompiling utilities or changing objects. Interpretation could take account of facts from the users environment (as is attempted by some ad hoc schemes in UNIX). File types would not have to be disjoint, and rules could be defined for arbitrarily complex commands, e.g. the comparison of two files of different types, for which some basis of comparison nevertheless exists.

4.4.4 Preserving filestore consistency

Certain properties of a UNIX filesystem are taken for granted by software both inside the kernel (e.g. for garbage collection) and outside (e.g. for archiving). Properties such as

"All children of a directory have distinct pathname components"

could be stated formally, and indeed phrased as Prolog questions to check whether, for any particular filestore, they hold. More urgently, attempts to modify the factbase (e.g. as files are created) must be thwarted if any of these assertions would be invalidated as a result. One obvious strategy is:

"tentatively make the change, then re-evaluate all assertions, undoing the change if any assertion fails."

A more practicable approach is:

"initially differentiate each assertion with respect to each type of update, assuming initial filesystem consistency, yielding a minimum check to be applied to each potential update."

For example, consider the creation of one new directory entry, and its effect on the property above. It is not necessary to recheck the unaffected directories, and verifying basename uniqueness within an affected directory of N entries is $O(N)$ rather than $O(NxN)$, since all previous names are distinct. Optimised checks such as these might be generated automatically from formal specifications of the filesystem. This topic is well-researched, if not conclusively so, within the realms of database theory.

4.5   OBSERVATIONS AND CONCLUSIONS

The following observations include comments on the Prolog language
and its environment, as well as more general statements on the use of a
knowledge-based approach to attribute management in a project support
environment.

4.5.1   The Need to Link Prolog to Other Languages

A consequence of wanting to use the right tools for each job is a
need to use Prolog in intimate conjunction with other programming
languages. There are two cases to consider.

4.5.5.1   Calling other languages from Prolog.   Consider the task of
filename expansion as performed by the shell. Prolog can readily evalu-
ate

        /usr/*/README

since the shell "wild card", in this instance, corresponds to an unin-
stantiated variable (although '.' and '..' are conventionally excluded
from the expansion). But

        /usr/m*/README

is more awkward, since Prolog does not provide regular expression matching as a primitive operation on atoms, and neither does it provide a
convenient basis for constructing it (unless pathname components are
represented as a list of characters). Since the algorithms have already
been efficiently implemented in conventional programming languages, it
is reasonable to want to repackage them as evaluable predicates in Prolog.

Another illustration of this need is given by the inadequacy of the
arithmetic facilities built into typical Prolog implementations. The
filestore application necessarily employs 16-bit unsigned integers to
represent inode numbers, and 32-bit unsigned integers for times (e.g. of
last modification to a file, as used by 'make' to determine
source/object consistency). One Prolog system, for PDP11 machines, supports only 15-bit integers, using the sixteenth bit of the machine word
for some private purpose.   Another system, for the PERQ workstation,
features twos-complement 29-bit integer arithmetic, and silently con-
verts larger integers to a low-precision floating-point representation.
These facilities may be better than nothing, but when applications have
a priori needs for standard precision, a means of linking in suitable
routines is essential.

4.5.5.2   Calling Prolog from other languages.   In the course of program
design, Prolog's data abstraction should be as readily employable as,
say, an array or a temporary file. Whenever some functionality seems
best expressed by knowledge, or whenever Prolog's pattern-matching seems
appropriate, a software engineer should be able to "declare" an instance
of Prolog. The EXPRESS project aims to provide an environment based on
data abstractions, in which such heterogeneity would be both supported
and controlled.

4.5.2   Concurrent inference

There are two aspects of this: Prolog programs contain implicit parallelism, which could be exploited to advantage on multiprocessor hardware. Conversely, when independent programs are run on a shared knowledge base, it will generally be unacceptable for the queries to be treated as atomic transactions, and serialised, since any may take unacceptably long to complete. Accommodating this "overlapping" parallelism is perhaps more important than accommodating the former.

4.5.3   Distributed factbases

In general, a (virtual) factbase might be composed of several scattered constituent factbases. Efficient inference under such circumstances might only be possible if the inference is steered to exploit whatever localisation of facts there is.

4.5.4   Redundancy and learning for improved performance

Traditional software techniques can achieve efficient handling of a limited range of operations by tuning the layout of data, perhaps including deliberate redundancy (such as the link counts within the UNIX filestore) to suit only the anticipated uses. While Prolog aims to support a wide range of queries. by representing knowledge canonically, there are opportunities for invisible internal duplication and indexing of facts to support particular queries, and thus give an improved mean performance when a strong pattern of repeatedly similar queries is made.

4.5.5   Binary Relational Form

Although Prolog supports facts of various length (and complexity), equal lengths are mandatory for matching facts to queries. So

        (mary , likes , cocoa , slightly).
        (mary , likes , whisky).

cannot both be matched by, for example

        ?-(mary , likes , X ...).

There is thus a class of rules that cannot be written, unless all facts are made the same length. The "binary relational" model has three elements per rule (a predicate and two arguments), and any fact, however long, can be transformed to a set of binary relational facts.

4.5.6   Negation as failure, or multi-valued logics?

In first-order predicate calculus, any statement that cannot be inferred from a factbase is considered to be false. So unless the factbase is complete, misleading results may be given.

More appropriate for practical applications would be a multi-valued system, with 'true' and 'false' recorded explicitly, and failure to infer being regarded as "don't know". Programming in a language based on such logic might not be as elegant as Prolog programming, nor as efficient, but these prices may have to be paid for serious manipulation of real data.

4.5.7 Non-monotonicity

Once a fact (or rule) is inserted in a Prolog factbase, it cannot be effectively withdrawn or denied by anything added subsequently. This is contrary to human reasoning, where overstatements such as

"if it!s a bird, then it can fly"

are refined by exceptions such as

"except that ostriches can't fly"

The concept of non-monotonicity[5] has not yet been resolved within the realms of theoretical logic, and is unlikely to appear in logic-based programming systems until that happens.

4.5.8 Why rules and facts?

Clearly, we have some faith that the fact/rule paradigm <u>can</u> make a significant contribution to the management of software development. The ability to generalise and abstract is a powerful component of human intelligence, and the statement of requirements in the form of rules seems to have promising applicability for building flexible systems. While "algorithms" and "data" interact deterministically in a way that makes them a feasible basis for useful computation, "rules" and "facts" interact according to the axioms and theorems of formal logic. The study described in this paper has provided sufficient evidence that a knowledge-based approach is a good way ahead for implementing tools such as configuration and version managers, and it is our intention next to write pilot versions of these tools.

ACKNOWLEDGEMENT

The financial support of the Science and Engineering Research Council is acknowledged.

REFERENCES

1.   D. M. Ritchie and K. Thompson, "The UNIX Time-sharing System," <u>Comm. ACM</u> Vol. 17(7), pp.365-375 (July 1974).

2.   W. F. Clocksin and C. S. Mellish, <u>Programming in Prolog</u>, Springer Verlag, Berlin (1981).

3.   S. I. Feldman, "Make - A Program for Maintaining Computer Programs," in <u>UNIX Programmer's Manual, Vol. 2 (Seventh Edition)</u>, Bell Laboratories (1979).

4.   O. P. Brereton, "Re-implementation of UNIX permission handling using Prolog," Report DCP/WD/164, Computer Science Department, University of Keele (April 1985).

5.   R. C. Moore, "Semantic Considerations in Non-monotonic Logic," <u>Artificial Intelligence</u> Vol. 25(1) (January 1985).

Chapter 5

Some UNIX tools to exploit a workstation

D.J. Barnes, J.D. Bovey, P.J. Brown and H.P. Siemon

UNIX† was designed in the days of teletypes, and its basic user interface remains unchanged to this day. It is now supplied on many personal workstations, which support a high resolution display, a mouse, and a windowing system. However UNIX in itself does not exploit any of these fine facilities. The only gain is that, if the user puts four windows on the screen, he has the equivalent of four teletypes instead of one.

This paper discusses some UNIX software tools that help exploit the full facilities of a graphics workstation. In terms of IPSEs or in terms of the Blit software (Pike (1)), they are modest tools: they would only form a small part of a programming environment, let alone a project support environment. Nevertheless they contain ideas — and salutary lessons — that may be valuable in a wider context, and this paper aims to bring these out.

Background

The tools have been developed by a team of four people at the University of Kent: the four authors of this paper. They have been implemented on the ICL Perq, under its PNX operating system (which is a port of Version 7 UNIX). Much of the work has been financed by the SERC under a grant whereby the University of Kent acts as a Software Tools Centre for the Common Base Programme. Because of the nature of the grant, there has been some emphasis on 'getting products out of the door' rather than long studies of what should be done. Clearly such an approach is not always the best one, but in this case we feel it has its merits. To quote a remark from a leading researcher at XEROX PARC: 'We still have no real idea of how to build software with good user interfaces; all we know is that most of our current interfaces are not good'. If you accept that in the short term — and we hope it is only in the short term — user interfaces need to be built in a largely empirical way, then there is some merit in getting tools out to the users as soon as possible, so that they can provide feed-back.

The nature of the tools

In this paper we shall discuss five tools:

- **oops**, a graphical debugger.
- utilities for displaying the UNIX file store.

---

† UNIX is a trademark of AT&T Bell Laboratories.

- **med**, a menu creation and editing system.
- **minit**, a window manager with a history mechanism.
- **guide**, a system for displaying documentation.

We shall describe each of these in turn.

Graphical debugger

The horrors of hexadecimal or octal core dumps are fading, though not fading as fast as we would like. We now expect symbolic, source level, dumps to be the norm. However, symbolic debugging systems still have problems: those systems whose action is to produce a dump of all current variables often produce so much information that the reader cannot find what he wants. On the other hand systems such as **sdb**, which require the user to hunt round a body of information, require much skill to use.

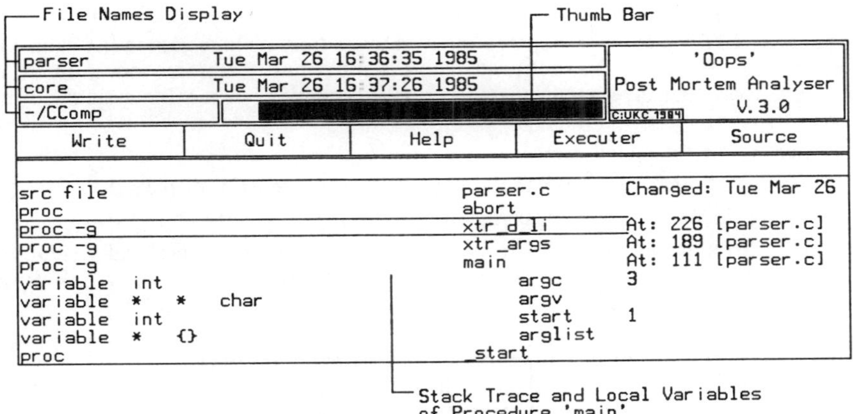

Figure 1: a display produced by the debugger

Oops, the debugger developed at the Software Tools Centre, finds a middle way. It gives the user a summary of the current state of the program when it produced its dump, and presents the user with simple methods of exploring the detail. These methods are based on the three following mechanisms:

- *pointing*: the user points at items he wants to know more about. Thus he might point at a procedure name that is currently active and he will be told its local variables.

- *scrolling*: large bodies of information are displayed in separate windows which the user can scroll through. Thus each array has its own window and the user can scroll through the values of its elements. Moreover the user can manage his screen, by moving windows about and by changing their size.

- *integration*: by using separate windows, the **oops** user can simultaneously see debugging information, the source program, results of a run and some 'help' information.

Oops has been designed to work with all the languages which exist on PNX on the Perq — currently Fortran 77, C and Pascal. This has been successful with Fortran 77 and C, but less so with Pascal. The problem is an instance of a common one: it is hard to implement an integrated system on top of existing diverse systems. Specifically the problem on the Perq is that Pascal is currently at odds with the general UNIX philosophy that is inherent in the C and Fortran 77 compilers.

There is, however, a more general and important point: a screen-based debugger that gives the user freedom to explore all the information about the current program state makes heavier demands on the symbol table than more limited debugging systems do. Thus it would be desirable to extend all the compilers to provide extra information in their symbol table. Similar considerations apply to the Blit debugger (Cargill (2), (3)).

In its current implementation, **oops** can be used only for post-mortem debugging. Future versions will, however, be extended to include more dynamic facilities. Figure 1 shows a sample of an **oops** display.

Utilities for displaying the UNIX file store

Nowhere is UNIX's teletype heritage more obvious than in its commands for dealing with the file system. The Software Tools Centre has therefore concentrated on providing tools that give a much more graphical interface to the file system. These allow

- walking the file system tree

- pointing at directory names to request that their contents be displayed

- editing file names and their permissions.

The aim, currently only partially realised, is to have a uniform graphical system that covers all the UNIX commands for manipulating the file system (**mv, cp, cd, ls, ln, chmod, rm, mount,** etc.).

```
┌──────────────┬──────────────┬──────────────┬──────────────────┐
│   page up    │  page down   │    help      │      quit        │
├──────────┬───┴────┬─────────┼───────┬──────┴────┬─────────────┤
│ cherish  │ remove │  move   │ copy  │ pathname  │  details    │
├──────────┴────────┴─────────┴───────┴───────────┴─────────────┤
│DIR rwxr-xr-x                                                   │
│DIR rwxr-xr-x          usr                                      │
│DIR rwxr-xr-x               demo                                │
│DIR rwxr-xr-x                    minit                          │
│FIL rw-r--r--                         clock.wa                  │
│FIL rwxr-xr-x                         minit_demo                │
│FIL rw-r--r--                         sh.wa                     │
│FIL rw-r--r--                         spy.wa                    │
│DIR rwxr-xr-x                    swim                           │
│FIL rw-r--r--                         clock.wa                  │
│FIL rw-r--r--                         sh.wa                     │
│FIL rw-r--r--                         spy.wa                    │
│FIL rwxr-xr-x                         swim_demo                 │
│                                                               │
└───────────────────────────────────────────────────────────────┘
```

Figure 2: a display of part of the file system

Figure 2 shows one of the file display tools in use. It can be seen that indentation is used to show the relative level in the file system hierarchy. The file that the user is currently operating on is highlighted. Otherwise the presentation makes modest use of graphics — we have instead aimed at providing a presentation that looks familiar to a UNIX user. We now feel, however, that presentation in many of our tools could benefit from the use of separate fonts. After all, the discrete use of different fonts on a printed page can lead to dramatic improvements in readability without any change of content. Similar considerations apply to the display of material on a screen: the above Figure might, for example, benefit if the directory names were in a different font from the other files.

Not only can tools display the standard UNIX file system, but they can also be used to collect other information about files and display it in a uniform way. For example our Perq software has an added facility whereby files can be *cherished*: cherished files are automatically archived every night and can therefore be restored when the Perq disk crashes. The 'cherished' attribute of a file can be presented to the user in much the same way as other file attributes, such as 'read permission', thus giving a natural and uniform interface.

Window manager with history

One of the innovations of Berkeley Unix was to provide a command history mechanism: the system remembers the last 20 or so shell commands that have been executed. The user can display this history at any time and selectively edit and re-execute commands from it. This is particularly useful in repetitive tasks such as an edit/compile/execute cycle.

However, the drawbacks of the glass teletype approach to history are several: screen space is limited and so it is usually necessary to redisplay the history each time it is required; finding the particular command required is often awkward; editing is cryptic.

These problems all but disappear on a multi-window workstation, and the **minit** system of the Software Tools Centre provides command history and window management in an integrated environment (4). Perq UNIX window management is done via a *window manager window* (WMW), which controls functions such as creation, deletion and size change of windows. The **minit** window manager also takes the approach that a user command to a window is passed through the WMW. All user commands are thereby retained as a single visible history in the window manager's window.

Users of multi-window workstations often pursue related tasks in different windows. The **minit** approach recognizes this and, to some extent, it alleviates the problems that arise in those systems which fail to recognize relationships between windows and regard each as a separate teletype.

Figure 3 illustrates the history as held in the WMW. A user can select any one of the command lines displayed, and, by using a pop-up menu, can apply any of the following operations to it: re-execute; delete from history list and edit; copy from history list and edit; re-execute in a new window (— the user is then asked to define the size and position of the new window); delete from the history list. Editing a command is a simple matter of selecting it, pointing at the characters to be changed and inserting or deleting as appropriate.

A menu at the top of the WMW provides options for the management of the history list. One of these options allows the list to be saved, so that it can be re-used in subsequent sessions; this facility provides a simple way for users to construct their own menu of frequently-used commands — the history list acts as this menu.

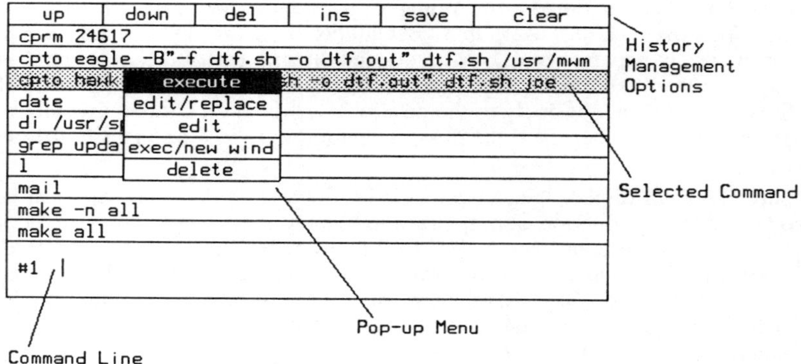

Figure 3: the window manager window

In summary, window management and command submission have been combined, allowing a flexible approach to history maintenance and command editing.

Menu creation and editing

Most screen-based tools interact with their users by means of menus. The user selects a menu item by pointing at it. In some cases the menu item may then present a subsidiary menu. (Thus a 'SEARCH' menu item may present a subsidiary menu containing the three items 'FORWARDS', 'BACKWARDS', and 'CIRCULAR'.)

People who implement screen-based software need an easy way whereby they can specify menus, and get feed-back when the user selects a menu item. Sometimes such facilities are deeply embedded in the operating system (e.g. Apple Macintosh software). This imposes a uniformity on applications. Arguably, a uniform user interface is the greatest asset a suite of application software can possess. System designers always underestimate the time that users will take to master the 'friendly' interface that has been provided, but if the interface fits a uniform style learning time is at least reduced.

Researchers, on the other hand, regard uniform user interfaces as premature standardisation. They wish to build new interfaces that overcome the inadequacies of current ones.

The **med** menu creation and editing system developed by the Tools Centre aims to tread a middle path between the uniform standard that the users want and the complete freedom that the researchers want. Implementors create and edit their menus by using a drawing package. The drawing package compiles menus into a form that can then be included in the implementor's programs. The system provides reasonable freedom for the implementor to use his own style of menu. There is flexibility in the style of menu supported (pop-up menus or static menus — where the latter may be within a window that changes dynamically in size), and in the type of feed-back provided when the user selects a menu item.

The menu drawing package has been constructed in such a way that the designer of the menu sees and uses his menu in practically the same way as the end-user of the final software in which the menu is to be embedded. The designer can thus experiment with his menu in a realistic environment.

The GUIDE document display system

Most computer-based systems for displaying documentation try to imitate the facilities available to a reader of a document on paper. Usually the imitation is poor and incomplete, with the result that users continue to prefer paper.

The GUIDE system aims to be a document display system that exploits the capabilities of a workstation and presents facilities that are impossible with paper-based documents. The GUIDE reader creates, from given source material, a document tailored to his own needs. A sample GUIDE session, dealing with source material about a photocopier might proceed thus:

(1) The reader is presented with a summary document which contains lines such as

FAULT REPORTING PROCEDURES   MORE

Items in bold, such as **MORE** above, are 'light buttons' that the user may point at.

(2) The user points at a light button, thus causing more information to be displayed. This material may itself contain further buttons. Buttons can be presented as alternatives, e.g.

Is the light at the side RED, YELLOW or BLUE?

(3) The user continues selecting buttons in any order he wishes. He can undo selections he does not want.

(4) If he wishes, he can edit the document simply by pointing at it and typing characters.

The process continues until the user has just the information he needs. If necessary the document can then be saved for further use. GUIDE also supports facilities for glossary-buttons, which allow users to create a glossary that explains terms they do not understand.

A prime aim of GUIDE is to make authorship of documents easy. This is achieved by an integrated approach whereby there is no firm distinction between authors and readers. In essence:

● a reader can edit his document, thus acting as author

● the *only* way an author can create a document is to explore the document as a reader would. There is no separate author language.

GUIDE can be used for documentation about any topic. It is valuable for documentation about computer software, particularly when that software is running in a separate window on the screen. (Indeed GUIDE provides the help system for **oops**.) Some macros have been written which automatically convert the UNIX manual pages to GUIDE format so that these can be explored in a more structured way.

Figure 4 shows a GUIDE window displaying an explanation of the UNIX **rm** command. Figure 5 shows the same window after the user has selected the **MORE** button that follows 'DESCRIPTION'.

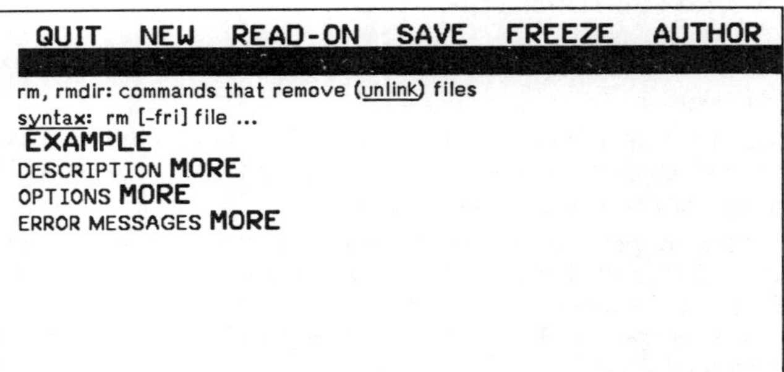

Figure 4: a GUIDE screen

*Figure 5: result of selecting a **MORE** button in the previous figure*

Criticism

The tools have been developed in a research environment, and the designers have to some extent explored their own ideas on user interfaces. There are common concepts shared by several of the tools such as menus, editing, scrolling, expansion/contraction (of the display of a procedure's variables in **oops**, of the contents of a directory in the file system utilities, and of buttons in GUIDE). These are not always presented to the user in a uniform way.

To return to a theme presented earlier in this paper, we often wonder whether it would have been better to devote, say, the first year of the grant to design studies and not to do any implementation until this was complete. Our general conclusion is that it would not have been better: the area is currently so poorly understood that we feel that our design studies would probably have focussed on the wrong topics and identified the wrong primitive operations. They may well have concentrated too much on one dimension of what is a multi-dimensional space. Instead it was much better to implement some tools and gain some user feed-back. Implementors must, however, have the courage to throw away the bad and the indifferent.

Conclusions

The five tools, all of which are still under development, constitute a small step to providing a friendlier environment for a UNIX user on a graphics workstation. All the tools escape from the teletype philosophy that you must master an elaborate command language before you use a tool. With our new screen-based tools, the user normally interacts by pointing at information displayed by the tool itself. The user needs to do little or no learning in advance. Several of the tools are exploring novel ideas; all of the tools aim at some form of integrated view. An ultimate aim is to give the user a uniform view of all the objects in the computer.

The saying "There ain't no such thing as a free lunch" has, as ever, made itself obvious to us. In our case the "lunch" is a "friendly mouse-based user interface". We have found that it costs a disproportionate amount of implementation time to refine and debug the way information is displayed. We therefore hope that, in the future, the outcome of projects such as DESCARTES (5) at Carnegie-Mellon and of the Alvey MMI initiative will at least make the lunch cheaper.

References

1. Pike, R., 1984, *Bell Labs. Tech. J.*, **63**, No. 8, 1607-1631.

2. Cargill, T. A., 1983, *J. of Systems and Software*, **3**, 277-284.

3. Cargill, T. A., 1985, *Software—Practice and Experience*, **15**, 153-168.

4. Barnes, D.J. and Bovey, J.D., 1985, *A Window Manager with Command History for the Perq*, University of Kent at Canterbury.

5. Shaw, M. *et al*, *Proc. ACM Symp. on Programming Language Issues in Software Systems*, 100-111.

Chapter 6

Some IPSE aspects of the Flex project

I.F. Currie

Introduction

Flex is primarily a capability based architecture [Currie et al. (1), Foster et al. (2)]. This architecture has been implemented in microcode and associated kernel software on several different hardware configurations, the most recent being the ICL Perq. Built on top of this as a normal unprivileged program is an object oriented operating system. It was designed to be used and programmed in a highly interactive manner. Perhaps the key-note phrase to be used in connection with Flex is "un-anticipated use" - one does know <u>exactly</u> what one is going to do.

The Flex system is the product of an evolutionary process based on the experience of its use and development over several years. I feel that many of the lessons learned in its construction could be usefully to an IPSE. A full description of existing Flex systems, or of the desirata of an IPSE are impracticable in this paper. There are just too many topics, ranging from the man-machine interface to the effective use of distributed processors, to do do all of them justice. However one important common area must be the support given to a programmer in the modularisation and reuse of programs, with its attendant difficulties in maintaining consistency between modules,texts and program. This paper is devoted to explaining some of the ways that Flex supports the programmer in this field.

Modules and program in Flex

Any large programming task must perforce resort to some kind of modularisation. Thus, the Ada compiler on Flex runs to about 200 modules while the operating system itself contains about 300. It is clearly crucial that the relationships between modules, texts and programs are easy to maintain. The following headings highlight some of the principal characteristics of modules in Flex.

●1 A module defines a piece of program which ,when obeyed, delivers some interface values that can be used elsewhere; the process of loading the module will obey this program so that the interface values are the <u>only</u> values which are visible. In Flex, the piece of program is usually a unit of compilation.

●2 Any program written in language L can use the interface values of a module provided that the types or modes of these interface values can be suitably mapped into L.

●3 An important instance of L in ●2 is the command interpreter language Curt. Any interface value produced by a module can be used directly by Curt; this implies that the mode structure of Curt is sufficiently general to encompass values produced by any module.

●4 The program associated with a module can be updated in such a way that the change is carried through to all programs which use the module, system wide. The right to update a module is given to the creator of the module and is transferable.

●5 No structurally inconsistent use of a module is permitted ie the use of a module remains valid if and only if the types and names of its interface values remains invariant. Any attempt at an invalid use is trapped.

●6 If the interface values of a module do change, then programs using it can be found easily and processed either by simply re-compiling their texts or by some other transformation.

●7 If the program associated with a module was produced by compiling some text, then the actual text used is derivable directly and unambiguously from the module.

●8 Even when using other modules, program texts can be independent of context. For example, module <u>values</u> can be included in the text, rather than their names (as after WITH in Ada). Taken with ●7 this implies that the complete tree of the texts of all modules used either implicitly or explicitly is available from the text of the program, regardless of the context of use.

●9 Any error reporting is dore in terms of the text(s) used to create the failing program so that one is always sure of the text of the program that one is diagnosing, including that of all of the modules used by it by virtue of ●8.

●10 All the previous versions of a module are potentially available and can be used to return the module to some previous state or to examine the sequence of changes to its text.

The Flex infrastructure

The Flex module system is implemented in the unprivileged part of the operating system. The infrastructure underlying this system consists of the Flex instruction set implemented in micro-code [Currie et al. (1)] and some kernel procedures [Currie et al. (3)] whose bodies operate in a privileged mode. The infrastructure is available to any user - he could replace the existing operating system with one of his own in just the same way as he writes more conventional programs. However the properties of the infrastructure are likely to strongly influence the kind of program that he writes, be it an operating system or not. This is because many of the facilities that it provides tend to be considered (if at all) at a much higher level in more conventional systems. This is paricularly true in the areas of security and integrity where the properties of capabilities have an immediate impact at all levels from the very lowest level of machine code upwards. Those parts of the infrastructure which are relevant to the module structure are described below.

¶1 Flex is a capability machine which treats capabilities as first class data objects. A capability gives the right to access some data or do some action if and only if one possesses the capability which cannot be forged. Saying that they are first class data objects means that they can be passed about quite freely in the same way as normal scalar data like integers or reals. Flex uses this capability structure to implement an object orientated system. These "objects" will usually be referred to as values and vary in type from simple integers or reals to much more complicated data or program structures containing capabilities.

¶2 The most important kind of value in Flex is the procedure. A procedure in Flex is implemented as a closure of code and values giving the non-locals of the procedure [landin (4)]. Amongst other things, procedures in Flex are used to enforce complex privacy relationships - they are simply general purpose capabilities whose properties can be programmed to produce any rights or restrictions.

¶3 Values are independent of the means used to produce them. For example, a value dug out of a failed program can be used with the same freedom as one produced more conventionally by calling a system procedure. This applies to all types of values from list structures to filestore dictionaries.

¶4 The underlying architecture gives considerable support in the treatment of exceptions. Raising an exception in Flex (either explicitly or implicitly by program error) means creating an exceptional value. These exceptional values can be trapped at any procedure call and can give a diagnostic chain of failed procedure frames and their code. Thus the locals and code of each level of failed procedure can be examined.

¶5 Filestore values exhibit the same generality of type and extent as other values One is not restricted to some fixed set of standard file types to implement data-bases but can use trees, procedures etc on filestore quite naturally since data written to filestore can include other filestore values. A simple kernel procedure call transforms a filestore capability to its mainstore analogue; with filestore procedures this kernel call is interpolated automatically on a call of the filestore procedure.

¶6 Most filestore values are 'write-once" values. They are created by writing some data to filestore using a kernel procedure and receiving back a capability to read that data, but not the capability to overwrite it.

¶7 Only a small number of root filestore values can be overwritten. The entire persistent filestore can be deduced by tracing from these roots. The only way the filestore can be permanently altered is by overwriting one of these roots with a different value. This operation is a unitary one and includes mechanisms to allow primitive controls for simultaneous access.

¶8 Provided that we arrange that root filestore values are independent of one another and never need to be updated together, the successive values in a root gives consistent snapshots of the data accessible from it. If we arrange to keep these values, a complete historical record of consistent states of the data is available. These states are consistent in the sense that the root gives either the old state or the new state - never some mixture of the two. This arises from the no overwriting rule in ¶6 and the unitary updating operation in ¶7.

Some implementation details

Although this paper is principally interested in that part of the system concerned with modules, any explanation of its implementation and use must widen the area of discourse. Thus, it has to include something about the Flex mode system, user environments and the Curt command language which covers rather a large area. However, this appears to be inevitable in any integrated system, regardless of the particular field. Here, then, is an overview of the implementation and operation of the Flex system where it impinges on the modularisation of programs.

§1   Flex implements a system-wide set of modes (or types) to describe the values that it manipulates. They were originally introduced in the Curt command language [Currie and Foster (5)] but now they pervade the entire system as a convenient method of describing the data representation and presentation. These Flex modes are somewhat more general than those of of ML [Gordon et al. (6)]. They include primitive modes like **Int**, **Real** or**Bool**,vectors, structures and references like **Vec Int**, **(Int,Vec Bool)**, **Ref(Vec Vec Int,Char)**. Procedure values are defined like **Int -> Real** , **(Vec Char, Int) -> (Ref Real, Int -> Real)**, etc.

§2   New atomic modes can be introduced to describe values whose structure does not require to be explicit. The mode **Edfile** is one such atomic mode which describes filestore values (each just a single filestore capability) which are generalised Flex text files. Values of mode **Edfile** are usually produced by the editor which has mode **Edfile -> Edfile**. **Edfiles** take full advantage of the fact that data written to filestore can include other filestore values. Thus, **Edfiles** differ from conventional text files met in other systems in that one can include values other than simple text characters in it. As a trivial example, a book might be an **Edfile** made up of a sequence of pairs of chapter headings and other **Edfiles** which contain the text of the chapters.

§3   An important extension of Flex modes upon ML modes is the infinite union described as **Moded**. Any Flex value can be represented as a **Moded**; this includes a representation of the mode of the value. It implies that objects like general purpose dictionaries can be implemented within the mode system. The mode **Vec Char -> Moded** could describe a procedure which searches a dictionary relating names to values of arbitrary mode . If a procedure, which delivers a **Moded** result, is called by the Curt command language, Curt will automatically "un-mode" its resulting value to give the underlying value with its underlying mode.

§4   In practice, the operation of Flex is divided into "user" environments. Each user is based on one filestore root (see ¶6). Making a new user involves the construction of a new name-space and a new module-space. A user is expressed as a filestore procedure binding its filestore root to the appropriate system actions so that it remains private to this procedure (see ¶2). Since all alterations to the user's name-space or module-space must be reflected in updates to the root, the history of all consistent states of the user can be remembered by keeping the old contents of the root when one updates it. Usually, this information gets thrown away by periodic filestore garbage collections (about once a week). There is, however, no reason other than limitations on size of physical filestore or archival medium why one should not keep complete history. A set of procedures allows one to query various aspects of this historical information, including modules as required by ●10.

§5   The name-space of a user is effectively defined by a partial function **find:Vec Char -> Moded**, belonging to the new user, which gives the value corresponding to its name parameter. The names acceptable to this function can arise either from a set common to all users or a set particular to the user over which he has control and can extend or contract. Among the set particular to the user are several concerned with creating and updating modules in his own module-space. Thus, the function **new:Compiledpair->Module** creates a new filestore value of mode **Module** from a **Compiledpair** value which is usually the result of a compilation (see §6). Similarly, the function **amend:Compiledpair->(Module->())** allows one to update a **Module** in this module-space provided that the type of its interface is unchanged so that ●5 is satisfied; **change_spec** with the same mode allows one to update it more drastically by changing this type. Possession of a **Module** value,whether in the creating environment or not, allows one to access its current **Compiledpair**.

§6   Compilers on Flex have the mode **Edfile->Moded.** The **Edfile** parameter is the program text while the underlying value corresponding to the **Moded** result is either a**Compiledpair**, if the text compiled correctly, or else an **Edfile** if the compiler found errors in it. In this latter case, the compiler calls up the editor on the offending text (indicating by the normal editing pointing mechanisms where the errors were found) to give one an opportunity of producing a correct text. Thus, getting a **Compiledpair** is usually the result of successive applications of the compiler which happens to include all the necessary editing to the original text. Note that this is the standard system editor called from within the compiler; both are standard unprivileged progams which can be used by any other program by any user. Indeed, most of the input to user programs is done by direct use of the editor from within user programs, giving a standardised method of interaction to all programs.

§7   The mode **Compiledpair** (whose name owes more to history than accurate counting) in fact consists of three filestore capabilities. The first component is a filestore procedure of mode **(Loader,Int) -> Keeps** (see §11) which is the actual compiled program. The second is a capability giving a codification of the "spec" of the program, ie the modes and names of its interface expressed in the language of the compiler used. {I use the abreviated term for this since it forms a small part of the actual specification of the program} The third component is the **Edfile** value which is the text which was used to compile this **Compiledpair** . Note that as filestore values cannot be overwritten (see ¶6), the correspondence between these three values will remain invariant. The relationship of the **Edfile** to the compiled program goes even deeper. Each procedure compiled will contain a filestore capability containing this **Edfile** together with diagnostic information relating the codeto text and local values to identifiers.

§8   If a program wishes to use the interface values of another compilation, it does so via a **Module** value derived from the **Compiledpair** resulting from that compilation. The text of a program which uses the interface values of a **Module** usually includes the **Module** value itself rather than its name in accordance with criterion ●8. The compiler makes use of the spec information to compile the correct code to use the interface values. Note that this code will only remain correct as long as the interface values have the types and names implied by this spec.

§9   Since a **Module** is effectively a variable containing a **Compiledpair**, the use of the **Module** rather than its **Compiledpair** allows one to update the program which produced the interface values without having to recompile the program using them in appropriate circumstances. These circumstances are specificallythose for which recompilation does not change the data in the spec part of the resulting **Compiledpair**, ie those for which the function **amend** (see §5) applies. This function will check to see that the specs given by the old and new **Compiledpairs** are the same and will leave the old spec capability in **Module** while updating the other two capability components of the **Compiledpair**.

§10 If it is necessary to use **change_spec** , then,at the very least, any program which uses the **Module** will have to be recompiled. It is clearly necessary that any attempt to use such a program without this recompilation must be manifestly invalid. The system achieves this by insisting that all compilers binds any **Module** used and the spec capability in its **Compiledpair**into its actual compiled program, ie into the filestore procedure which is the first component of the **Compiledpair** (see §7). Usually these values will be the only non-local values bound into the closure which forms this filestore procedure. Before the compiled program is loaded, this spec capability remembered at compile time will be compared with the one currently in the **Module**; unless they are identical, no loading will be permitted.

§11 The compiled program (see §7) is a procedure , and contrary to the usual practice, we apply the compiled program to the loader rather than vice-versa. The mode of this compiled program is **(Loader,Int) -> Keeps**. The **Int** parameter is not relevant here. The **Loader** parameter is a procedure which will be used by the compiled program to apply to any **Module** <u>directly</u> used and hence bound with its compile-time spec (see §10). The compiled program can choose how and when this call is done and is not constrained to any particular format by system fiat; this is the principal reason for the inversion of general practice mentioned above. The **Keeps** answer to the procedure is a mainstore capability to a block containing the interface values as defined by the particular language being used. The representation of these values can always be expressed in terms of the Flex mode system.

§12 A **Loader** procedure has mode **(Module,Spec)->Keeps** and is applied to each internal**Module** used by the compiled program itself. The other parameter is the spec capability of this **Module** bound with it at compile time. This loading procedure performs three main tasks. First, it ensures that its **Module** parameter still possesses the spec given by its **Spec** parameter. Secondly, it remembers those **Modules** to which it has been already applied to avoid evaluating the same interface values more than once. Thirdly, provided that this is the first application to a given **Module**, it evaluates that **Module**'s compiled program (whose mode is **(Loader,Int) -> Keeps**, remember) with itself as the parameter and delivers its **Keeps** result as the result of the **Loader** . Thus the complete program formed by the graph of the use of **Modules** is loaded and evaluated by a single call of the compiled program procedure with an appropriate **Loader** parameter.

§13 Since the interface values obtained by loading a compiled program have a Flex mode, they can immediately be expressed as values in the Curt command language. Curt is a very simple reverse-Polish language which can only construct parameters to procedures, call them (provided their modes are correct) and, possibly, name the results. As an example of this kind of transformation suppose that we had an Ada package text in an **Edfile** called **ada_file** which defined two procedures, push and pop, operating a stack of integers. In orderto apply the Ada compiler to it, load the program

and name the resulting procedures with two Curt identifiers, the
following Curt line is sufficient:

> **ada_text ada ! run_ada ! = (push,pop)**
>> {the ! symbol says apply proc to previous element}
>
> where
>> **ada:Edfile ->Moded** is the Ada compiler (see §6)
>>
>> **run_ada:Compiledpair->Moded** loads the compiled program ,
>> evaluates it to give the interface values and uses its Ada
>> spec to deduce its Flex mode. The Flex mode in this case
>> would be **(Int->( ), ( )->Int)**.

After obeying this Curt line, the Curt identifiers **push** and **pop**
can be used in simple tests of the Ada program from the Curt line
like:

> **42 push**!

and check that the value of:

> **( ) pop**!

isindeed 42.

§13 The use of "foreign" language modules by a compiler for some
language L will require that the foreign language interface values
can be mapped into language L. Clearly this will sometimes be
impossible; for example there is no analogue in either Pascal or Ada
for an Algol68 procedure which delivers a procedure as an answer.
We know in Flex that all interface values can be described by Flex
modes; thus the rule to see whether language specs in different
languages are interchangeable is simply to see whether their
mappings into Flex mode system are identical. The identifiers used
to name these values are to a large extent irrelevant. The particular
forms of the identifiers may depend on the language and their use
is only of interest to the particular compiler.

§14 An exception arising from a program error will usually be trapped at
the Curt level at the call of a procedure. This call will produce an
exceptional result of mode **Exception** which can be analysed in a
variety of ways. The most usual method of diagnosis involves
applying the procedure **diagnose:Exception->Void** to the exception.
This will allow one to display the sequence of calls which produced
the error by pointing at text used to compile the procedure (using
the diagnostic information given in the code of each procedure, §7)
and giving access to the local values of these procedure calls as
ordinary Curt values. These include all kinds of values eg locally
declared procedures could be called in the same way as **ada, run_ada**
, **push** and **pop** were called in §12.

Conclusion

This explanation of the Flex module system includes quite a lot of detail in many aspects of the system. Throughout it, however, phrases common in other system descriptions like "name of" , "indentifier for" or "hierarchic control" hardly figure at all. This gives the clue to the essence of Flex which is that Flex deals directly with values rather than deriving then via some names. Further, these values are global in extent – one does not have to be in a particular context in some hierarchy to use them. This is somewhat foreign to most systems which can only deal directly with the most trivial of values. For example, most systems only allow access to filestore via some filename held in some directory. Flex, on the other hand , allows one to deal directly with filestore objects. One can create them by writing data to filestore and keep them in any way one pleases, unrestricted by system defined names or directories. This immediately alleviates problems concerned with ensuring that one is in the correct context for interpreting the names or making sure that their meanings do not change. Named objects in filestore play a very small part the Flex system – most user directories contain fewer than twenty named objects.

The Flex module system is elegant and simple to use precisely because one can treat modules and the other objects involved directly as values, by-passing unnecessary forays into naming regimes. I believe that most areas in an IPSE would benefit both in use and implementation from the same treatment.

References

1. Currie I.F., Edwards P.W. and Foster J.M. 1981 "Flex Firmware" RSRE Report 81009.

2. Foster J.M., Currie I.F. and Edwards P.W., 1982 " Flex: a working computer based on procedure values", Proc. International workshop on high-level architectures" Fort Lauderdale, Florida.

3. Currie I.F., Edwards P.W. and Foster J.M. 1981 "Kernel and system procedures in Flex" RSRE memorandum 3626.

4. Landin P.J. 1964 "The mechanical evaluation of expressions" Computer Journal Vol 6 No 4 pp 308 - 320.

5. Currie I.F.and Foster J.M. 1982 "Curt: the command interpreter for Flex" RSRE memorandum 3522.

6. Gordon M.J., Milner A.J. and Wadsworth C.P. 1979 "Edinburgh LCF" Springer-Verlag (Berlin)

Chapter 7

An overview of the ASPECT architecture

J.A. Hall, P. Hitchcock and R. Took

7.1   INTRODUCTION

ASPECT is an Integrated Project Support Environment which, working on a distributed host, will support software development for distributed target systems. The ASPECT project is supported by the Alvey Directorate and is being carried out by Systems Designers, the Universities of York and Newcastle Upon Tyne, Mari, GEC and ICL.

ASPECT will support all phases and activities during software development. Naturally this includes programming, in a variety of languages, but it also encompasses specification and design in both graphical and textual form, as well as management, planning, integration, testing and maintenance.

In order to provide this wide ranging support, ASPECT must meet two key goals. The first is that it should be an open environment : open to new tools, new methods, new languages, new hosts and targets and new methods of user interaction. The second is that it must provide true integration of all ASPECT - supported tools.

These twin goals are reflected in a major objective of the early stages of the project : to define an architecture for the ASPECT system which provides both openness and integration. This paper outlines the ASPECT architecture. Section 2 gives an overview of the whole architecture, while sections 3 and 4 discuss in more detail two major ASPECT components. Section 5 summarises and draws some conclusions.

7.2   ARCHITECTURAL OVERVIEW

The ASPECT approach to integration is not to impose a single framework - for example a particular development method - for that would contradict our policy of openness. Rather, ASPECT offers to tools a powerful set of common services for structuring and storing information, for communicating with users and other tools, and for manipulating remote targets. This set of services is made available to tools through the public tool interface (PTI) and it is the software offering the PTI which is the heart of ASPECT. To achieve the required openness the PTI is extensible, to support new methods and tools, and configurable, so a project can impose particular methods of working, if required. Furthermore, since the PTI is the only means by which tools use ASPECT services, it necessarily includes within itself the facilities for its own extension and configuration.

One of the most important requirements on ASPECT is that existing tools, written for the host operating system, should be usable within ASPECT. This is made possible by the open tool interface (OTI). The OTI can be thought of as a subset cf the PTI which appears to the tool just like the host operating system. ASPECT is hosted on UNIX*, so the OTI makes available to ASPECT a large collection of existing software development tools.

The PTI services fall into four groups:
- information storage
- man-machine interface
- process invocation and communication
- target services

Each of these services is provided by a layer of software - in effect a subroutine library - between the tool and the UNIX Kernel. The PTI offers services at a much higher level than those of UNIX. To provide the open tool interface, the PTI includes calls appearing to be UNIX system calls, but even these are processed by ASPECT rather than by UNIX so that all tools, including UNIX tools, are fully under the control of ASPECT. Indeed, different open tool interfaces will be provided for different UNIX tools to capture the semantics of the tools' data correctly in the ASPECT information base.

Underlying the ASPECT code is a distributed UNIX system. This is built, on different and geographically distributed host computers, using the Newcastle Connection (Brownbridge et al. (1)). This is a particularly powerful and elegant technique which makes a set of distributed UNIX systems look like a single UNIX.

The overall architecture of ASPECT is, therefore, as shown in Figure 7.1.

Two clases of tools are shown. Perspective (an existing Systems Designers product) and ASPECT Tools use the high level services of the PTI ; UNIX tools use the OTI facilities within the PTI. The public tool interface, shown as a double line, provides access to the ASPECT services : the user interface, information storage and tool invocation (both included in 'ASPECT database') and the target interface. Underlying these services are the various host machines, all running UNIX, distributed over wide area or local networks.

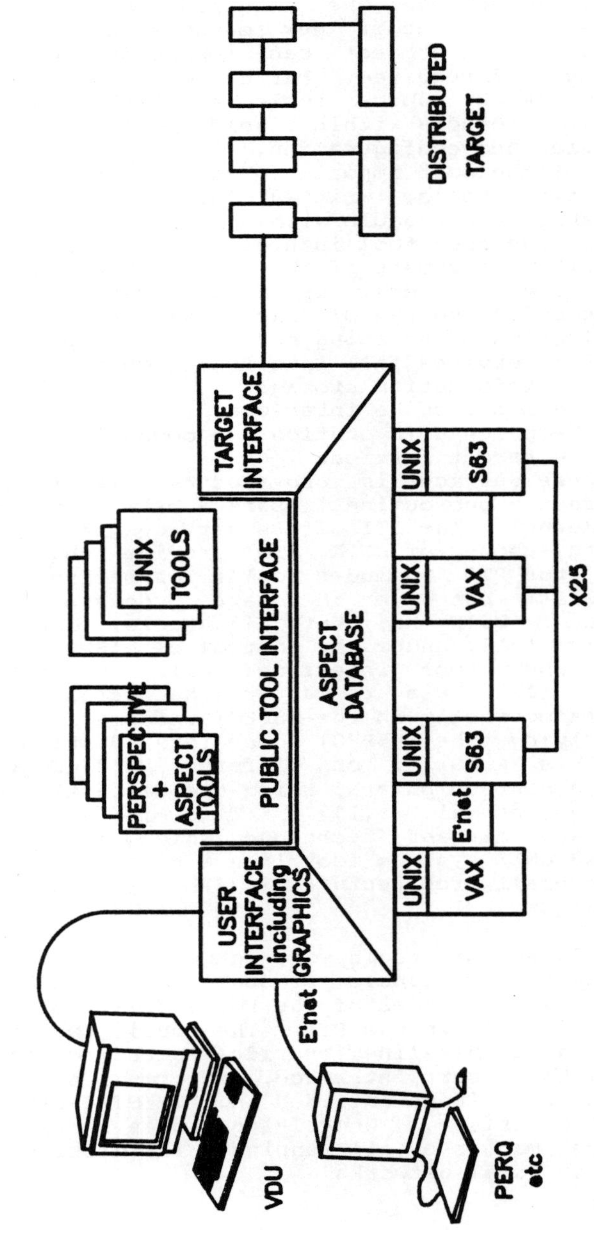

Fig. 7.1   Overall ASPECT Architecture

7.3 THE INFORMATION BASE

7.3.1 Integration and the ASPECT Database.

THE ASPECT architecture depends on a central database as the main way of achieving integration between tools. In a purely file based environment, each tool contains the data structure definitions that it needs to interpret the records which it reads and writes. If two tools have been designed to work together, they might share the declarations for these data structures via a common library. Database takes this notion one step further by insisting that any such data declarations are held centrally and describe one neutral view of the data which is then made accessible to everyone. Every database system uses some standard data definition language (ddl) and also provides a standard way of manipulating the data via a standard data manipulation language (dml). Data produced by one tool will then be in a standard form and readily available to others. A view mechanism allows the standard view to be transformed where necessary for the requirements of other tools.

The overall architecture of the database follows the model of ANSI/X3/SPARC (Tsichritzis and Klug (2)). Three levels are defined with the view mechanisms defined as mappings between them. (Fig 7.2).

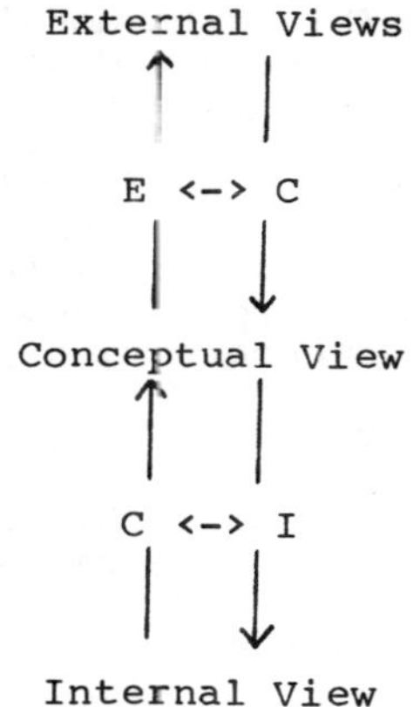

Fig 7.2 The ANSI/X3/SPARC Architecture

The central, or conceptual, level is where our standard view of the whole application domain is defined. This is done in an implementation independent way. The internal level and its associated mapping allows the conceptual model to be mapped onto an implementation. It is at this level that we will cope with distribution. At the conceptual level, there will be a single system view as far as possible, mapped onto implementations on a number of machines at the internal level.

The external level and the mapping from the conceptual to the external level provides the view definition mechanism. This allows different subsets of the conceptual model to be presented to tools in the ways that they require. UNIX views of the data will be given for the OTI.

This architecture has the advantage that the conceptual model can be extended as requirements change, whilst preserving to a large extent the original external views. Only the mapping needs to be redefined. No tools should need to be rewritten. The other possibility is that the implementation can change. New access paths might be defined, indexes added, or distributed data moved from machine to machine. Again tools at the external level are insulated from these changes by the appropriate mappings.

It is not enough, of course, just to use database technology. There must be disciplined usage of the database, normally achieved through the offices of a Data Base Administrator. The data model for the software development process must be agreed at the conceptual level so that it can serve all expected tools and be easily extended to hold the information required by tools not yet thought of. It this is done, and tools are not allowed to define their data in isolation from other tools, then all the advantages that a database system has over a pure filing system will be realized. These advantages are:

<u>reduced redundancy</u> : Unnecessary duplication of data is avoided.

<u>standard languages</u> : Any data combination is now avaiable to any tool via the standard ddl and dml.

<u>multiple views</u> : Tools see only the subset of the database that they need, presented in the way which they require.

<u>data independence</u> : tools are shielded from representation details. In some sense, the database is a sophisticated abstract data type.

<u>security</u> : Security can now be handled in a centralised manner rather than in an ad hoc manner by each tool.

<u>recovery</u> : Again this is handled in a centralised manner rather than by each tool. Central to this is the notion of a transaction. This is an active unit of work which is either carried out completely or, if aborted, returns the database to its original state.

<u>integrity</u> : It is now possible to centralise rules about permitted values and transitions of the database, and to

ensure that tools do not transgress them. This is particularly important when access to the information in the database is opened up to many new tools by the OTI and the PTI.

7.3.2 The Conceptual Data Model.

Central to the database architecture is the conceptual level. We have chosen to use the language RM/T (Codd (3)) to express the conceptual model of ASPECT.

RM/T is an extended form of the relational model for data. The extensions allow for better definition of the semantics, or integrity rules, of the application domain being modelled than does the pure relational model. RM/T also gives better support for view definition.

We chose RM/T for the following reasons:

First, the pure relational view of data can be kept for the majority of database use. This view is simple, well defined and understood, and is supported by a general query language.

Second, at the conceptual level RM/T provides a modelling language within which an entity relationship model of the application domain can be expressed. It offers an advantage over the pure relational model in that some integrity rules, particularly referential integrity, are built into the model, thus removing one of the major problems associated with the pure relational model. We see these integrity rules, together with those that are specific to software engineering, as a vital part of the database system. It is these rules that will ultimately ensure that the database remains under the control of ASPECT, even when being accessed or modified by arbitrary sets of tools.

Third, stable relational database systems are commercially available as implementation vehicles at the internal level. A relationally based model such as RM/T will map easily onto such an implementation. Relations offer another advantage at the internal level: they provide a closed set of high level operations. This is very useful in a distributed environment. A high level request can be sent to a remote database and a relation, containing precisely the data needed, can be returned. This relation can then be easily fitted into the subsequent evaluation mechanism.

Last, an integral part of the RM/T model is its catalogue. This is essentially an integrated data dictionary which is an important part of any database system, particularly if it is to be extensible. Further, the catalogue is recursive, in that it will allow some objects to have the dual role of being both data and meta-data. This is an important charactristic of objects found in the software engineering domain.

7.3.3 Some Relations for the Software Engineering Domain.

We will sketch here the important relations we see as being part of the conceptual model for the software

engineering domain.   Further   details may   be   found   in
Hitchcock et al. (4).

RM/T has the notion of <u>surrogates</u>.   Every object   has
its own internal system identifier.   The relation <u>known-as</u>
may be used to give these objects names within the context
of a name space in a way exactly analogous to UNIX files
and directories.   An   object   may be anonymous and
recognised by other properties or may have several
external names in different name spaces. This will allow
access from different workspaces, a characteristic of the
software   engineering   process.   Names   may   also be
contextual, relative to a particular name space, or full
path names relative to a base name space.   This is
particularly useful when naming information, such as make
lists, is to be passed from context to context.

It is another characteristic of the application area
that objects may have several types as their roles change.
Something developed as a piece of program text might later
become a tool.   Types also must be considered as objects
which can be related to each other in a type hierarchy.
These concepts are directly captured in the RM/T model.

Versions of objects can be considered to be instances
of some higher order object type, which might itself be an
instance of a yet higher order object.   In this manner any
arbitrary version hierarchy can be modelled.

Our paradigm for the software development process is
that it consists of a sequence of   representations,
starting with a requirement specification and ending with
executable code.   Each representation is transformed by
some process, or script, into   the next, and each
representation is verified to be consistent and to model
the semantics of the previous representation (McDermid and
Ripken (5)).   The scripts may be carried out by humans
(e.g. design) or be purely mechanical (e.g. compilation).
This paradigm is represented in our model by three
relations. A given script <u>requires</u> certain objects as
resources, <u>uses</u> certain objects as input, and <u>produces</u>
other objects as output. The notion is highly recursive;
scripts can themselves be produced from smaller ones using
the appropriate scripts for script combination.

We have used the rather small set of relations
mentioned above to describe the UNIX tools of SCCS
(Rochkind (6)), MAKE (Feldman (7)) and BUILD (Erikson and
Pellegin (8)), and in fact to enhance them, so that a
history of the production process is kept.

Finally, we have to consider the <u>granularity</u> of the
database. We have been deliberately non-committal on this
point. At one extreme the database could be seen as a
sophisticated directory into the UNIX filing system, whose
use is made mandatory by trapping the appropriate UNIX
system calls.   At the other extreme, much more of the
primary data might be held directly in the database.
Exactly where to draw the line will depend on particular
circumstances. We hope to be able to support a spectrum
of choices.

7.4 THE MAN-MACHINE INTERFACE

7.4.1 Design Considerations.

The architecture for ASPECT's man-machine interface is derived from four fundamental premises. First, ASPECT will accomodate a range of tools. Tools incorporated under ASPECT's Public Tool Interface will comprise both tools designed specifically for ASPECT, and tools adapted to run using ASPECT resources, while tools imported within ASPECT's Open Tool Interface will be largely unchanged from their original operating environment. Second, ASPECT will support a range of terminal devices, principally from vt100-style devices at one extreme to bitmapped, mouse-driven workstations like the PERQ at the other. Third, ASPECT will serve a range of users. Users will vary, for instance, in their role within ASPECT – from project managers to programmers – and in the degree of their competence and familiarity with computer interaction. These two factors imply a range of styles of interaction: managers may prefer a menu-based dialogue, while experienced programmers may prefer more command-driven interaction. Last, it is recognised that within a particular style and on a particular terminal device, a uniformity or at least consistency of interaction over the range of tools is a prerequisite for an efficient and credible man-machine interface to the ASPECT system.
These premises have immediate consequences for the architecture of the ASPECT MMI. Their most critical feature is the openness of the ranges: not only must ASPECT cope with new users, but also with new tools and possibly new terminal devices. Handling such open ranges requires a means of representation abstracted from particular members of the range to a level independent of them, and at the same time comprehensive enough to allow the inclusion of new members. The openness of the ranges also makes it impossible to specify a definitive configuration of tools, users, and devices as a basis for design.
The design of the ASPECT MMI architecture has therefore concentrated on the organisation of these different representation into a unified model with generic components capable of handling any particular configuration. Principally, the architecture allows for the decoupling of the components of the interactive system at three points along the continuum between the user/terminal and the software tool. First, in order to mask the specific hardware characteristics of particular terminal devices from the rest of the system, a level of device independence must be provided above which input/output specifications share a common representation over the range of terminals. Second, in order to assimilate the interactive requirements of particular tools into a unified interactive 'style' common to the whole range of tools, a level of tool independence must be provided above which, in addition to the I/O

input/output, and in terms of the dynamics of the interaction. The IM is parameterised by a style specification which configures this communication into, for instance, a menu- or command-driven mode. Each tool is decomposed into a set of functions and an Interaction Specification written in a custom-built high-level language. The IM interprets the Interaction Specification and calls the tool functions appropriately. Tool output is passed by the IM to a Virtual Display, one or more of which are associated with each tool. The Virtual Display is an intermediate data structure which stores a generalised description of the state of the screen for a particular tool or tool function.

Each terminal device has associated with it a software module which handles the mechanics of interaction. In the case of a vt100, this is a device driver which maps a particular Virtual Display onto the screen, possibly with some graphics emulation. In the case of a bitmapped screen, the controlling software module is a window manager which, under user control, presents each open Virtual Display in a window. Input from either device is first filtered by its associated module which performs echoing and string-packaging. The window manager in addition tracks the mouse cursor and appropriates user input relevant to the control of windows. The remaining processed input is passed back to the IM to be treated as command or data in accordance with the current interaction state of the relevant tool.

7.4.3   The Internal Interfaces.

7.4.3.1 Device independence. This is achieved via the Virtual Display for output, and a canonical protocol for input. The Virtual Display will be a data structure incorporating both graphic and textual information in a normalised coordinate schema. Input protocol will be standard.

7.4.3.2 Tool independence. This is achieved by means of an Interaction Specification Language. This will enable the semantics, or the 'deep structure' of the interaction of each tool, to be encoded for interpretation by the IM. The raw input/output between tools and the IM will be standard.

7.4.3.3 Style independence. Style specification is effected by means of a parameterisation mechanism on the IM. The IM accepts a restricted set of style parameters governing such factors as display style and verbosity. The style specification thus determines the syntax of the interaction.

7.4.4   Features and Problems.

A consequence of the modularisation and decoupling effected by the model via the interfaces is a clear representation, there is a common representation for the dynamics of the interaction. Finally, the ability to select from a range of styles and impose the selected style on subsequent interaction necessitates a level of style independence above which exist generic functions common to all ASPECT systems. The ASPECT architecture attempts to embody this structure as a hierarchy of levels of independence, from specific device characteristics at the base of the tree, up to a generic control module at the top.

Transitions between levels require precisely defined interfaces which essentially map a more abstract description onto a specific description via an intermediate representation, in one direction supplying any information hidden at the higher level, and in the other direction filtering out such information. Thus the interface implementing device independence will on one side deal in terms of sequences of control characters specific to particular terminals, while on the other will recognise abstract representations of terminals embodied in logical devices.

A number of criteria for the effectiveness of these interfaces can be isolated. The degree of exploitation of the domain onto which they will map will be important. For example, ASPECT's terminal devices range from bitmapped screens to vt100s. It is important that the capabilities of these devices be used to the full – it would clearly not be satisfactory simply to emulate vt100 performance on a bitmapped screen. Secondly,the more abstract representation employed by the interface should be complete and consistent, and should not restrict the functionality to be communicated via the interface. Lastly, it is important that any intermediate representations be economical and efficient.

This approach to structuring the MMI architecture has the advantage not only of fulfilling the initial premises, but also of leading to a highly modularised system, in which incremental or differential changes to tools or styles can easily be made since dependencies are few and strictly defined. If decoupling of system components via interfaces in this way were not adopted, sensitivity to specific characteristics of modules and devices would be propagated throughout the system, such that tools, for instance, would have to be aware of terminal characteristics and stylistic options. This would lead not only to longer, more complex and more hidebound code, but also to the breakdown of the central authority over interaction and consequent stylistic anarchy.

7.4.2 The MMI Architectural Model.

The model is laid out to reflect both flow of control and data and the hierarchy of levels (fig. 7.3).

The Interface Manager (IM) is a generic component common to all ASPECT IPSEs. It manages the communication between the tools and the user both in terms of basic

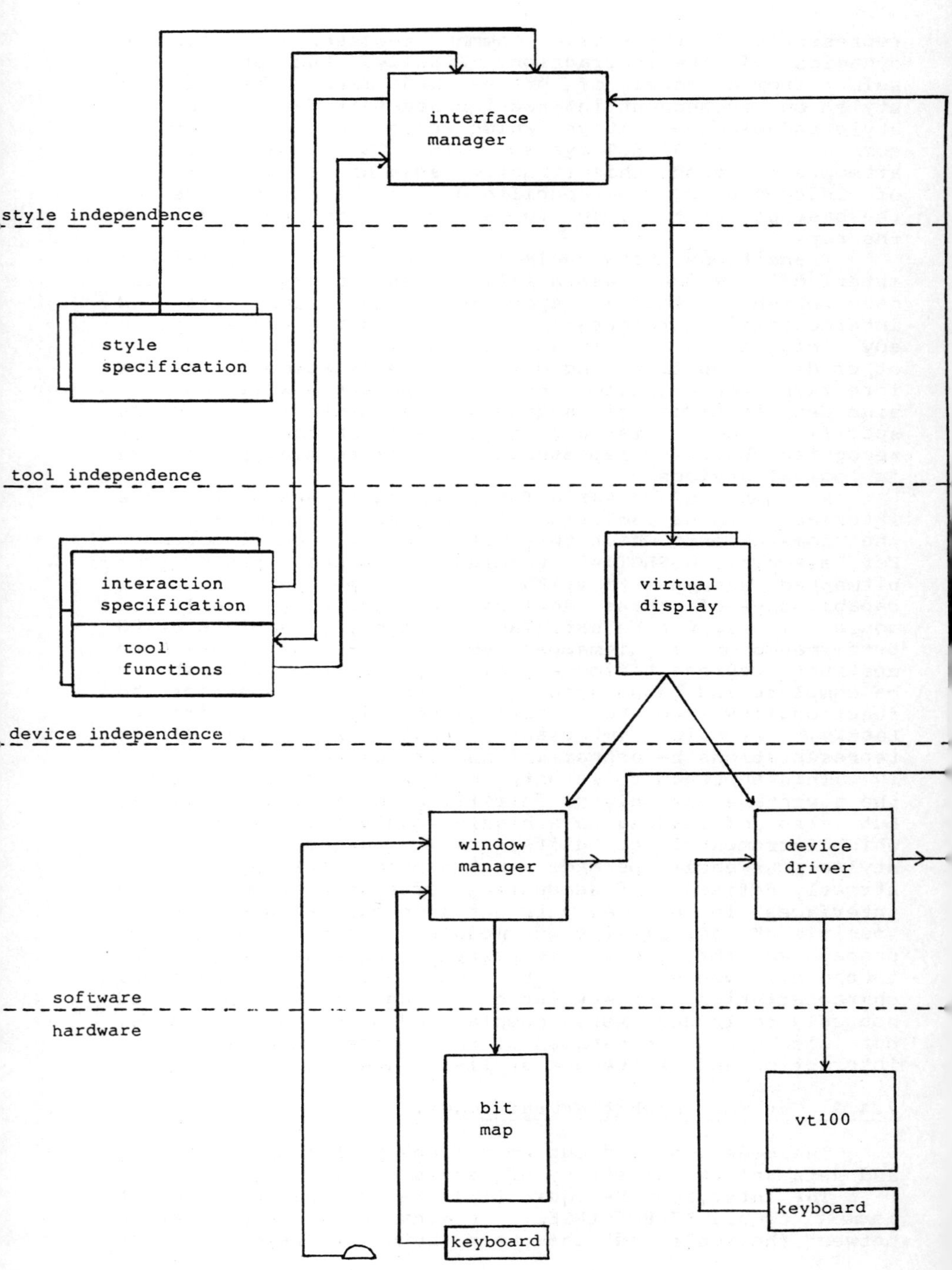

Fig. 7.3   ASPECT MMI Architecture

demarcation between domains of user control and domains of tool control. Tools have jurisdiction, via the IM, over the existence and contents of the Virtual Displays, but the user has control over the configuration of the windows presenting each Virtual Display.

The IM itself may initiate a Virtual Display, for instance when it requires to display a menu for selection by the user. In this case it will create a Virtual Display containing the menu, and this will then have display priority, to be presented either as a pop-up menu on a bitmapped screen, or as a full screen menu on the vt1ØØ. In addition, the overall command language can itself be treated as a tool, and thus be subject to stylistic variation in harmony with the other tools in the environment.

A number of problematic areas can be identified within this model. The performance of the architecture is likely to be critically dependent on the efficiency of the various internal interfaces. Most of the interfaces are 'channel' types, but the Virtual Displays are 'pools'. This may lead to problems in incremental screen updating. One solution may be to provide a bypass channel directly from the IM to the device. Another might be to bring highly interactive functions like screen editing down to the device dependent level. In both cases the Virtual Display should be kept up to date in parallel so that redrawing of the screen is possible.

A further problem is the status of tools imported under the OTI, since these will not be expected to provide a separate Interaction Specification. This argues for the the provision of a default Interaction Specification, which allows the tool autonomy to handle its own interaction. In these case, of course, the style of interaction cannot be guaranteed to conform with that set in the environment.

Three types of man-machine interface can be identified. A fixed interface has a single style of interaction. An adaptive interface monitors user performance and modifies the style of interaction accordingly. The ASPECT MMI belongs to a third type, the tailored interface, in which the user may select his style of interaction. Apart from this characterisation, the architecture presented here is an attempt to construct a mechanism capable of generating a wide range of styles of interaction whilst minimising the effect of tool or device dependencies. In this way it is hoped that the MMIs to ASPECT IPSEs can be configured and fine-tuned iteratively in fast response to user requirements and preferences.

7.5 SUMMARY AND CONCLUSIONS.

The twin goals of integration and openness are fundamental to the ASPECT architecture. Integration is achieved by providing a unified set of common services to manage all ASPECT data and interactions. Openness is achieved by the extensibility of these common service and in particular by the ability to support, within this framework, existing (UNIX) tools. We believe that the architecture we have presented in a solid basis for a flexible and open-ended IPSE.

*UNIX is a trademark of AT & T Bell Laboratories.

REFERENCES

1. Brownbridge, D.R., Marshall, L.F. and Randell, B., 1982, <u>Software - Practice and Experience</u>, <u>12</u>. 1147 - 1162

2. Tsichritzis, D.C. and Klug, A. (eds), 1978, The ANSI/X3/SPARC DBMS Framework: Report of the Study Group on Data Base Management Systems', <u>Information Systems</u>, <u>3</u>.

3. Codd,E.F., 1979, <u>ACM Transactions on Database Systems</u>, <u>4</u>, 397-434

4. Hitchcock, P., Whittington, R.P. and Robinson, D.S, 1985, 'Modelling Primitives for a Software Engineering Database', <u>Proceedings British National Conference on Databases</u>, to be published.

5. McDermid, J.A. and Ripken, K., 1984, 'Life Cycle Support in the Ada Environment', Cambridge University Press, England.

6. Rochkind, M.J., 1975, <u>IEEE Transactions on Software Engineering</u>, <u>SE-1</u>, 364-370.

7. Feldman, S.I., 1979, <u>Software - Practice and Experience</u>, <u>9</u>, 255-265

8. Erickson, V.B., and Pellegrin, J.F., 1984, <u>Bell Labs. Technical Journal</u>, <u>63</u>, 1049-1059.

Chapter 8

An overview of the ECLIPSE project

A. Alderson, M.F. Bott and M.E. Falla

8.1 INTRODUCTION

Eclipse is an integrated project support environment which is being developed by a consortium led by Software Sciences Ltd and supported by the UK Alvey Programme. The other members of the consortium are CAP (UK) Ltd., Learmonth and Burchett Management Systems, the Universities of Lancaster and Strathclyde and the University College of Wales, Aberystwyth.

This paper presents the background thinking which led to the Eclipse development and gives an overview of Eclipse itself, and of our plans.

8.2 THE REQUIREMENT

The major theme of the Eclipse Integrated Project Support Environment (IPSE) is productivity in the context of large projects.

We believe that among the most important factors for raising productivity are:

* automation
* a well-structured approach to the work
* effective use of brain power
* training and continuing guidance
* re-use of existing components

The approach that one takes to these factors depends critically on the size of project in question. The techniques appropriate to a team of one or two people working for a few months are totally different to those appropriate to a team of fifty or a hundred working for two or three years. Eclipse sets out to address the larger end of the size spectrum and the activities of organisations which typically have a number of projects in progress at any one time.

The problems which large projects bring with them are:

* problems of complexity and control
* the involvement of many people with consequent communication and management problems
* the probable use of several computers possibly dispersed on several sites
* diversity in the methods, standards and languages used between projects, and even within a project

In setting out to design an IPSE to achieve high productivity on large projects three opposing and conflicting pressures have to be resolved. The first is that the amount of work required to produce a truly integrated project support environment, together with a full range of tools for all the relevant methodologies and languages, would be colossal.

The second pressure is that of time. The world will not wait while such a comprehensive IPSE development is undertaken and run to completion. Certainly British industry requires at least a modest IPSE well before it could be completed.

The third pressure is that of the future. The problems of scale and timing could be solved by acquiring and bolting together in some haphazard fashion a more-or-less related collection of more-or-less relevant software tools. The result would have been scarcely satisfactory in the present, and would offer no kind of base for building into the future.

Our response to these countervailing pressures is the concept of an 'open', evolutionary and permissive IPSE. We have set ourselves the target of devising a basic structure with its own coherent philosophy, which is nevertheless capable of supporting the mainstream development methodologies, notations and languages. At the same time we have required of our design that we should be able to use, mostly without change, all the tools which are available under the host operating system.

8.3 THE ECLIPSE PHILOSOPHY

Our philosophy is based on the view of the system development process which is illustrated in figure 8.1.

We start at top left with the user or customer who has some need or desire for a computer system. This need is typically vague and only loosely specified. The process of system development can be regarded as a series of transformations from one representation to another, each

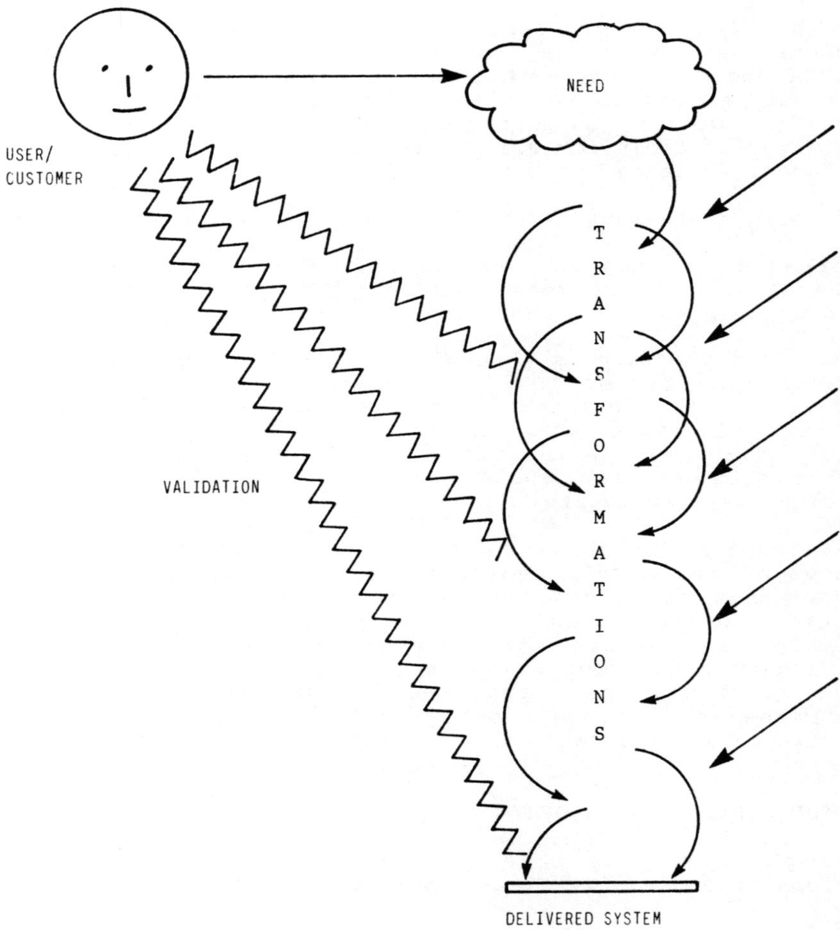

Figure 8.1   A view of the development process

stage introducing greater precision and greater rigour. The final system, at the bottom of the diagram, inevitably involves total precision and total rigour (even if not total correctness).

Most stages in the chain of transformations involve the incorporation of new knowledge. In the early stages one is investigating the application environment in order to fill out the expressed need, e.g. for 'a stock control system', into a complete specification. At later stages one is injecting information about the hardware that is available and the constraints which it imposes. At each stage decisions are being made.

The arrows indicating the transformations have deliberately been shown as overlapping. This is because different methodologies involve different stages and different numbers of stages. They also involve different representations and a different stance on the question as to whether it is better to use the same representation or distinct representations at each stage.

What we have been able to observe is that the different representations used in the different methodologies make similar demands on the underlying support system. At a superficial level, many are graphical. At a deeper level, all involve version and variant handling, most involve the 'composed of' relationship, the useful ones have rules governing what is legal and meaningful. When applied to developments of any size all the methodologies need extra support for project control purposes.

An IPSE must therefore be able to record entities and their relationships. It must be able to cause work to be done and to record the results and history of this work. It should also monitor the rules which apply to the relationships between entities, to the entities themselves and to the development process.

In figure 8.1 we have deliberately refrained from labelling the various stages or phases of system development to emphasize that an IPSE must not be committed to a particular partitioning of the total process. In traditional terms however we could label the column of transformations with terms which would run something like: feasibility study, requirement specification, outline design, detailed design, code. We also note that the term 'delivered system' includes user documentation, system documentation, test data and results, and so on.

Of course, system development is never a matter of uniform and continuous advance from stage to stage. It would be naive to expect that the user, having expressed his need and possibly supplied some supplementary information as input to transformations, will be happy with the resulting system. At several stages in the development we seek to check back with the user that he is satisfied, on the basis of the current representation, with the system which he will get. This is the process of validation and it is indicated by the ziz-zag lines in figure 8.1.

System development, however, is not a one-off process. In practice, the world changes, the user's needs change and the available hardware changes. We wish to avoid starting again when a change is needed.

At present, a common problem is that only the last representation, the system code, is modified and earlier representations (design documents, specifications, etc) may be allowed to go out of date. Even when earlier representations are updated, there is rarely any automatic checking of this, and the reasons for the changes are rarely recorded.

An IPSE should, at the least, facilitate the task of updating all representations by chaining together, in both forward and backward directions, related items in each representation.

An IPSE should also, however, assist in the rapid recapitulation of the development process. Any system change will be provoked by either a change in user need or a change in the information which has been injected into the transformations at some stage in the development process. Ideally it should be possible to recapitulate the entire development process, actually reworking only those parts of the transformations which are different.

The product of a development project is therefore not just code, nor even code plus documentation and test data, it is a 'derivation structure' which records how the system was derived from needs, observations and assumptions, and which allows similar systems to be re-derived.

The key to our philosophy therefore is:

*   a succession of system representations, expressed in a variety of textual and diagrammatic languages.

*   a control environment in which these representations
    can be recorded and related to each other in complex
    ways both within and across stages in system
    development.

   Within this broad philosophy, methodologies can be seen
as:

*   defining the language or diagrammatic notation to be
    used at each stage
*   giving rules, or at least hints, to guide the
    transformations from stage to stage

   Methodologies differ in how many stages they specify, in
how much detailed guidance they give on how to proceed
from one stage to the next, in what parts of the total
development process they cover, and in the rigour with
which they treat the process. One needs also to remember
that many organisations work, for at least part of the
development process, with informal or loosely specified
methods.

   In the Eclipse project we have adopted a two-level
approach to the support of methodologies. At the first
level we are providing general purpose mechanisms for
recording, editing and analysing both text and diagrams of
the sort widely used in a range of methodologies. Users
or advocates of particular methodologies will be free to
use these facilities and to build tools which interface to
our standard representations.

   For certain methodologies the Eclipse project will
implement a range of specific tools. We have chosen to do
this for LSDM (a version of which is widely used within
Government under the name SSADM) and MASCOT. This will
serve to check that our representation techniques are
correct and to demonstrate to others what can be done. We
call this degree of incorporation 'level two'.

   Clearly we wish our tools to be as general purpose as
possible. To this end they need to work on standard
representations of data. For text data, we have a
standard: ASCII coding in UNIX files. For the
representation of entity-relationship-attribute data we
have introduced a standard based on the Interface
Definition Language (IDL) developed at Carnegie-Mellon
University (1) which we call IDLE (IDL for Eclipse).

8.4 OBJECT MANAGEMENT

Tools form one component of the drive towards the automation which was identified as desirable in section 8.2. The other major component is support in handling the very large number of revisions and variants of items of all kinds (documentation, specification, source code, binary, test data, plans, schedules, progress data, etc), and assistance in organising the construction of products (macro processing, compiling, building, text formatting, etc). This is provided by two subsystems within Eclipse that have been inherited from Software Sciences' own Foundation development: the Object Management subsystem and the Execution Management subsystem.

The Object Management subsystem uses the SDS2 database management system (2) to record **objects**, each of which consists of control attributes (which are SDS2 database fields) and a body (which contains the user's information in the form of text, binary data, structured data in a form defined by IDLE or indeed in any other form). A minimal set of control attributes for objects is included in the basic schema; it can, and normally will be, extended to meet the needs of a project and of specific tools. The bodies of objects are never updated.

The naming scheme is hierarchical; nodes in the tree are **catalogues** and the leaves are **items** and **indexes**. Objects are normally recorded as **versions** of items, but objects may remain unattached to any item. Eclipse does not impose any particular numbering scheme for revisions and variants: the user project can define its own scheme for organising the versions of each item into a system of revisions and variants.

Although the naming tree is the relatively stable, long term, project-wide view of the structure of the database, each team and each user normally accesses the database through one or more indexes which allow a user to set up his own names to identify objects, items, catalogues, and indeed indexes. This then reflects the team's or the user's view of which of these things are important at the present time. Each user, whether human or in the form of another subsystem, sees the database through a particular **context**. From within a context, a search path, containing catalogues and indexes, is used to find the appropriate interpretation of a name.

Objects are produced by executing **transformation procedures** (themselves stored as objects), which are defined in terms of the UNIX commands; they are the means by which existing software tools are incorporated into Eclipse. The writing of transformation procedures is regarded as a privileged activity to be undertaken by system programming staff, since such procedures must not compromise the integrity of the database. The ordinary user may define **derivation procedures** which consist of invocations of transformation and of other derivation procedures, together with direct operations on the database and the ability to stop processing if defined conditions are not met. A derivation procedure may specify the derivation of a single simple object (e.g. a single compilation) or the complete build of a large system or subsystem involving dozens of operations and several final products.

When an object is created, all the details of its derivation are recorded in a **derivation record** associated with the object. This record can be used

* to recreate the object from its origins: this often permits space to be saved by removing object bodies from storage
* to create similar, but slightly different, objects: thus one can request the derivation procedure which would have built an object, and then by editing that procedure, or by running it in a context with a different index, one can obtain a new object which incorporates different versions of components or different components.
* by the SDS2 query language to analyse the origins and destinations of components, subsystems, releases, etc.; this allows, among other things, the impact of proposed changes to be determined.

Taking this idea of derivation as something recorded in the database a stage further, Eclipse allows us to specify a 'derivation plan' which records all the transformations necessary for the build of a system (or a document) in advance of any of the components actually being available. This gives us the benefits of the database and its query language at the early planning stage.

The derivation record also helps to record how elements in successive representations of a system development are related to each other, thereby allowing us to record and re-use a complete system derivation, as described above in section 8.3.

8.5 THE USER INTERFACE

In section 8.2 we identified the effective use of brainpower as an important factor in obtaining high productivity. Using appropriate methodologies obviously reduces the problems of complexity, and suitable automatic

tools and object management save time and mental effort.
Beyond that we need an effective man-machine interface
which will

*   be efficient and attractive to use
*   provide power-assistance to the brain of the user who
    is attempting to grapple with the difficult problems
    of system development

In order to support fast response to user interaction,
at least while he is performing highly interactive tasks
such as text and diagram editing, the principal user
interface to Eclipse is based on a high-resolution
bit-mapped workstation with 2Mbytes of RAM and 40Mbytes of
hard disc linked by Ethernet to a central computer. The
workstation supports multi-processing with window
management, allowing the user, for example, to consult
and/or edit several documents at once.

We expect a range of tools to be developed, both by us
and by others, to run within the Eclipse environment
partly or wholly on the workstation. A standard is
therefore being defined for tool writers which will offer
significant support in the production of attractive human
interfaces.

Although workstations of this type are still relatively
uncommon in everyday use, we are conscious that hardware
technology is moving rapidly in this area and that
advances in screen sizes (or, equivalently, resolution)
and the introduction of grey scale and colour at practical
costs are not far away. We therefore plan an interface
which will carry tools forward unchanged into this
enhanced environment through the use of concepts such as
virtual fonts and colours.

The means by which we provide 'power-assistance' to the
brain of the user is still a matter for research. Clearly
not being hung up waiting for trivial operations helps, as
does a coherent organisation of data and systematic work
plans. Much work has been done in other places ((3), (4),
(5)) on coherent and uniform user interfaces to the whole
range of tools provided. While we recognise that this is
important, our policy of incorporating existing tools,
including for example, existing text editors, puts a
severe limit on what we can do here. We should, however,
be able to develop a coherent 'metaphor' or user model of
the system which places the user in a simplified
intellectual environment in which his tasks, resources and
products become visible and manageable. This metaphor
will then be supported by the user interface and the tools
which the user employs.

8.6 REVIEW

We are now in a position to review the contribution we are making to the productivity factors and problem areas which we identified in section 8.2. We have outlined above our approach to automation, the effective use of brain power, the management of complexity and the diversity which we want to support. Here we turn our attention to the other factors and problems.

8.6.1 A well-structured approach The Eclipse project is not doing any new methodological work. Instead we plan to support a wide range of existing and yet-to-be-developed methodologies. We have identified LSDM and MASCOT for major attention, partly because of their practical use and partly because they, and LSDM in particular, cover a wide range of techniques. We also intend to incorporate, at least at level one (by the parameterisation of standard Eclipse facilities) VDM, JSP, JSD and possibly others.

Although the derivation procedure mechanism outlined above gives us the capability to define a system structure in advance of code components, we regard this only as a first step along the road leading to a fully non-procedural System Specification Language. Research towards this goal is therefore a component of the Eclipse project.

8.6.2 Training and continuing guidance The task of training and continuing guidance in Software Engineering is one which the Eclipse project itself cannot, of course, address (although several of the participating organisations are active in this field). We are, however, committed to a design policy which, while aimed at professional users, will minimise Eclipse-specific training and the need for re-familiarisation by intermittent, occasional or specialised users.

8.6.3 Re-use of existing components The most potent aid to productivity is the re-use of existing components. Effective re-use, however, is hard to achieve and requires suitable design and interfacing techniques, a catalogue and retrieval mechanism based on a sound organising principle, and a powerful mechanism for glueing together components which may have been written in different languages without necessarily much attention being paid to their ultimate re-use. Nor is it possible to consider re-use in vacuo; experience has shown that it is necessary to identify a particular application area in order to constrain the options.

We have therefore set out to address these problems in the context of fast prototyping of command and control systems. This application area has the advantage of supporting other Eclipse aims, such as contributing to the validation of early stages of system development. It also builds on the application expertise of several of the collaborating organisations.

8.6.4 **The involvement of many people** Eclipse will offer the project management facilities in SDS2, which include a critical path analyser for hierarchically structured project plans. In addition it will be possible to record and analyse multiple versions (i.e. revisions and variants) of plans.

Project managers will also be able to use the Object Management subsystem to monitor the real status of system components. Thus, for example, once an item has achieved 'clean-compiled' status this can be detected, and even linked to planning information, automatically. The programmer who has declared a module 'complete', and who has achieved clean compilation for it, cannot subsequently tinker with it or make it 'really complete' without management's knowledge.

8.6.5 **The use of many computers** The use of several computers comes about in three different ways

*   using workstations
*   host-target working
*   multiple hosts

We have seen above how we intend to distribute Eclipse over a local area network connecting many workstations to a central host. This host, a VAX computer, will hold the central database and perform the bulk of computation-bound work, leaving detailed interaction with the human user to the workstation.

We envisage Eclipse being used quite frequently in a host-target mode where it is used as an environment for supporting development work, with the option of running tests and the final application in some other environment. This second environment could be a large mainframe or a small micro. In order to blaze this trail we have elected to provide a host-target development environment for systems to run on the Intel 80286, programmed in Ada, and capable of being developed using the MASCOT methodology. A compiler, linker, builder, loader, run-time system and test controller will be provided, together with support within the Object Management subsystem for Ada program libraries (6).

Since we are directing our efforts at large developments we must cater for projects which will need more than one host. Initially the user will be aware that several instances of Eclipse are involved, but with very convenient transfer of data between them. We plan to move towards a situation where the user sees one unified IPSE executing on computers that may be separated by a few yards or many miles.

8.7 DEVELOPMENT STRATEGY

Eclipse is being built around two proprietary products developed by Software Sciences Ltd. The first is SDS2, a database management system designed specifically to support software development by large teams (2). It is the succesor to SDS ((7), (8), (9)) which dates back to 1971 and which has been used on a number of large development projects.

SDS2 employs an entity-relationship-attribute model with bi-directional links, facilitating the representation of the complex network of relationships characteristic of software development. A wide variety of data types is provided, including user-defined enumeration types, variable length text, sets of links, and subrecords. SDS2 supports an experimental update capability and very long transactions; it uses optimistic concurrency to by-pass the problem of deadly embrace in a multi-user environment. Its query language and interactive browse capability will be available through Eclipse.

Foundation, the second contributing product, is an extension to SDS2 which addresses the problems of object management and the controlled production of derived objects (such as system builds). Foundation forms an integral part of Eclipse, and the description in section 8.4 above essentially describes Foundation.

Our strategy has therefore been to base Eclipse upon Foundation. However, whereas SDS2 and Foundation have been developed as products, with all that implies in terms of completeness, testing and documentation, Eclipse is being developed as an engineering prototype, that is as a vehicle for trying out the ideas and their usefulness, but possibly lacking a certain degree of 'finish'. It is part of the Eclipse plan to make interim releases for evaluation by ourselves and others.

We have clearly not started with a clean slate and our policy is to strike a balance between a clear understanding of the design issues and options implicit in

the IPSE concept, on the one hand, and putting together a
practical system which will give real experience of these
issues within a limited timescale and budget on the other
hand. The various subprojects comprising the Eclipse
project   therefore   range   from   straightforward
implementation at one end of the spectrum through to
relatively pure research at the other.

8.8 CONCLUSION

We believe that in Eclipse we are producing not only a
sound engineering prototype, but a foundation for much
future research and development.

Within the current project we have some effort set aside
to consider future developments and we conclude this paper
with an indication of where we expect substantial advances
to be made by ourselves and others over the next few
years. These are:

* pushing the formalisation of the development process
  as a series of controlled transformations forward into
  the early feasibility and requirements phases through
  the use of IKBS techniques
* tool integration to match the data integration which
  Eclipse   provides;   this   will   be   achieved   by
  establishing an execution environment which supports
  highly   communicative   but   independently   executable
  functional components.
* organising system documentation differently e.g. as a
  network of concepts, tied to the formal parts of the
  system derivation, rather than as a linear sequence of
  paragraphs; this will reduce the cost and increase the
  accessibility and usefulness of documentation
* incorporating   intelligence   into   the   user   interface,
  firstly by implementing rules to guide and to check,
  subsequently to offer 'secretarial' support to the
  user; eventually we envisage knowledge about the
  system   derivation   process   being   integrated   with
  application knowledge in a knowledge base and being
  manipulated by IKBS techniques.
* the application of formal methods to the verification
  of individual transformation steps, and to the task of
  checking the final system back against the early
  statement of requirements (coupled with 'common-sense'
  world knowledge) by generating suitable test data and
  prescribed results.

The task of developing a first generation IPSE within
acknowledged historical and time constraints is generating
an increasingly clear perception of what a second
generation IPSE, built on a green-field site, would look
like.

ACKNOWLEDGEMENTS

Although most of the ideas and concepts expressed in this paper have been the work of the present authors the Eclipse project is now some 60 strong. The carrying forward and evaluation of these ideas into a working system must be placed to the credit of our colleagues.

We also wish to acknowledge with thanks the contribution of RSRE to the development of SDS, a forerunner to SDS2, of the NCC Software Products Scheme to the development of SDS2 and the VMS version of Foundation, and of the Alvey Directorate to the Eclipse project.

REFERENCES

1.   Nestor J R, Wulf W A and Lamb D A 1981, 'IDL-Interface Description Language: Formal Description', CMU-CS-81-139, Carnegie-Mellon University.

2.   Software Sciences Ltd., 1984, 'SDS2 Overview'.

3.   Thacker C P, McCreight E M, Lampson B W, Sproull R F, and Boggs D R, 1979, 'Alto: A Personal Computer', Xerox PARC CSL-79-11, Palo Alto.

4.   Adele Goldber, Dave Robson, 1983, 'Smalltalk-80: The Language and its Implementation', Addison-Wesley.

5.   Teitelman W, 1984, 'The Cedar Programming Environment: A Midterm Report and Examination', Xerox PARC CSL-83-11, Palo Alto.

6.   Pierce R H, 1985, 'Ada in the Eclipse Project Support Environment', Proceedings of the International Ada Conference, Paris, Cambridge University Press.

7.   Falla M E, Burns D, 1973, Datafair Proceedings, 116-173, 'Software Development Systems'.

8.   Software Sciences Ltd., 1982, 'An Introduction to SDS'.

9.   Software Sciences Ltd., 1972, 'Software Development System'.

Chapter 9

SPRAC: From specifications to programs

J. Foisseau, R. Jacquart, M. Lemaitre, M. Lemoine and G. Zanon

<u>9.1 BRIEF OVERVIEW OF SPRAC</u>

SPRAC (3) is an Integrated Computer Assisted Software Development System. It interacts with the user for the expression of formal specifications, for the purpose of producing and validating software components. SPRAC is concerned with the development of classical programs, e.g. programs written in imperative languages such as PASCAL or ADA.

SPRAC is mainly concerned with the formal specification of abstract data types (ADT) and functions (F), and the definition of algorithms (ALG).

<u>9.1.1 The main characteristics of SPRAC</u>

The main characteristics of SPRAC are the following:
1) the ability to describe the software objects at three distinct levels, corresponding to specification levels more and more close to the computer.
2) the existence of a project data base (PDB) which presents a structured working area. The schema of the PDB is relational.
3) the existence of a knowledge data base (KB) from which "software components" can be searched and extracted.

At the level of design languages, we find an assertional language, called LF, which is based on the first order logic. LF permits specification of functions, abstract data types, and representation of abstract data types. All these objects can be generic. The genericity is very close to the one used in ADA.

The utilisation of LF requires a method of modular specification: objects are independent and developed one at each time.

At the second level, LA is an applicative language used to describe algorithms.

These two languages in conjunction with the PDB are macroscopic tools used in software production.

The last level, not yet included in SPRAC, is called the LM level, and therefore corresponds to the level of programming languages. In practice all software produced with the help of SPRAC will have an operational level in LM.

It is hoped that the target language for LM will be a subset
of ADA.

9.1.2 Explanations of some choices

Why three levels of languages?
It has been decided to introduce three levels of
languages in order to isolate the problems which are usually
treated simultaneously and in a single language. Indeed
these problems are:

1) what are the abstractions used to solve a given
problem?
At the LF level, representation and control are not
taken into account. Only the logical properties of a
possible solution are expressed. At that stage of development, we are not concerned by the non functional properties
of the objects.

2) what is the best suited control?
LA is used to describe HOW to get a result. Using LA,
we control the sequence of statements. Globally LA permits
the description of algorithms with functional behaviour
according to the classical control structures.
There exist local variables to which values can be
assigned. The parameters are transmitted by value: therefore, no side effects are permitted in LA. An algorithm
imports only functions and/or ADTs. This guarantees the
independance of the representation of the ADT. We have to
distinguish two classes of algorithms:
- the actual algorithm which corresponds to functions.
- the "algorithm type" which corresponds to a set of
  algorithms, each one corresponding to the algorithmic
  expression of an operator of an abstract data type.

3) assuming the preceeding points, what are the best
data structures?
We think the link between an ADT and a concrete data
type must be supported by the PDB. Indeed, there are
several representations for a given ADT; the choice of the
representation is a development step and that choice must be
preserved by the PDB. The description of the representation
will be made first of all in LF, then in LA for the
algorithm part.

4) According to the choices made for the algorithms and
the data, how do we allocate the memory of the computer?
The LM level will allow this expression.
LF and LA enable the design and realisation of
independant software objects. The communications between
them are described through interfaces.
These languages have been designed and implemented to
facilitate quick specifications and algorithmic descriptions
with little regard to performance. They allow the production of a first prototype in a rather small amount of time.

This prototype can be considered as the starting point of
the following steps of software production.

9.1.3 The Project Data Base and its role in the software process

The project base concept for software engineering in
SPRAC comes mainly from works on automatic program synthesis
(6),(5).   It extends program synthesis to "programming in
the large" with active assistance to the user. The project
base structure was designed in order to take into account
what is left implicit in program synthesis: non-functional
properties.
The kernel of SPRAC, the PDB, permits the definition
and the management (checking and realisation) of static and
dynamic links between the objects described at the three
possible levels.   These links are established according to
two dimensions: SPACE and TIME.

1) space dimension.
The objects are linked by relations such as USE.
REALISE, REPRESENT, INSTANTIATE.
All the objects belong to an entity called a VERSION
which corresponds to snapshots taken during the development.

2) time dimension.
To keep track of the way people develop software. the
genealogy of the versions are preserved in the PDB, as a
tree of versions: the history.   All the versions and their
contents (objects and links) are frozen.
The next sections focus on one aspect of SPRAC: how a
given methodology of software development can be taken into
account. In other words, how the (not so well known) laws
of software development can be used within a support
environment such as SPRAC.

9.2 LINGUISTIC MEANS TO EXPRESS DESIGN KNOWLEDGE WITHIN SPRAC

The linguistic means used in order to assist and to
model the software development process are mainly those of a
Data Description Language and Data Manipulation Language of
a relational data base management system with deductive
capabilities (8).
The result of software development is not only the
final produced programs but also a structured description of
the whole development.
The Conceptual Schema of the Project Data Base is
intended to be a model of the software development process.
The kind of capabilities used to define and implement this
Conceptual Schema are:
- a development framework to describe partial solutions, to
  keep track of what is being done and what has been done,
  etc.
- languages -at least partially formal- in order to describe

objects included in partial solutions.
- deductive capabilities used to assist actively -or even
  directly- the users.

9.2.1 Objects of the development, object base structure

A project base consists of a collection of development
states and general rules for static and dynamic integrity
constraints.

9.2.1.1 - Development state structure. A development state
is a structure consisting of:
- an intermediate solution,
- a local agenda, which indicates what could or must be done
  to reach the next development state from the current one,
- management data relative to the intermediate solution.

An intermediate solution is built from usual "programming"
objects. Each object is itself a structure each component
of which we call a property. There are four kinds of
properties: definitions, operational, validation and management. The three last properties are named
"non-functional properties".

A - List of the objects
As it has been written in the first paragraph. the objects
supported by the current implementation of the PDB are:
- functions, used to specify input-output relationships.
- data types, seen as a collection of related functions (so
  called operators)
- algorithms, normally linked with functions
- data type representation,
- algorithmic definition of data type representation.

B - Linguistic levels for the definition of the objects

i) 'Semantic' level for functions, data types and representations
Functions, data types, data type representations are defined
by means of LF. One of the main design criteria of this
language was that the LF specifications had to be
executable. Each specification consists of four parts:
- a header, identifying the object,
- requirements, which state the hypotheses on the environment: some data types or functions with specified
  properties have to exist.
- an interface, giving what can be seen from the outside,
- a sequence of defining formulae: formal specification in
  predicate logic of input-output relationships for functions or operators of abstract data types.

ii) 'Semantic' level for algorithms and algorithmic defini-
tion of type
These objects are defined by a typed higher order language,
LA. The data types used in these algorithmic definitions

are those of the LF level (the "concrete" data types are expressed in LF).
Each algorithmic definition consists of four parts. The first three parts are common with LF specifications. The last one is a sequence of formulae, each of them defines a function by algorithmic construct (conditional, recursion, application, lambda-abstraction).

iii) 'Syntactic' level of description
All the objects previously described are represented and handled by means of abstract syntax trees. These trees are themselves coded by Lisp S-expressions.

Exemple: the following formula defining "set inclusion":

    incl(S1,S2) <=> ALL x:item, isin(x,S1) => isin(x,S2)

is represented in LF by:

```
(EQUIV (incl S1 S2)
       (OR (AND (ALL ((x item))
                     (IMPLY (AND (isin x S1))
                            (AND (isin x S2)))))))
```

An LF formula is basically a disjunction of conjunctions. The syntax used for the denotation of function application is the standard Lisp syntax.

C - Other properties of the objects
These properties have been previously called "non function-al" properties. They are:
- validation properties; depending on the kind of objects, are memorised test data sets, verification conditions, mathematical properties, etc.
- management properties; for example date, author. creation time, etc.
- identification properties; these properties are used for data base retrieval, reuse of knowledge, etc.

All these functional and non functional properties are represented, stored and handled as clauses of a prolog-like data base for retrieval purposes.

D - Structure of an intermediate solution
An intermediate solution is a set of objects related to each other. The relationships among them are:
- a "realise" relation between algorithm and function,
- a "use" relation between functions, data types, etc.,
- a "represent" relation between data types and representation functions,
- an "instantiate" relation between a generic object and each of its instances.
An intermediae solution is a "frozen version" or a copy of a "frozen version". Other entities are related to the evolution of versions. The PDB permits the definition and

management of temporal links.

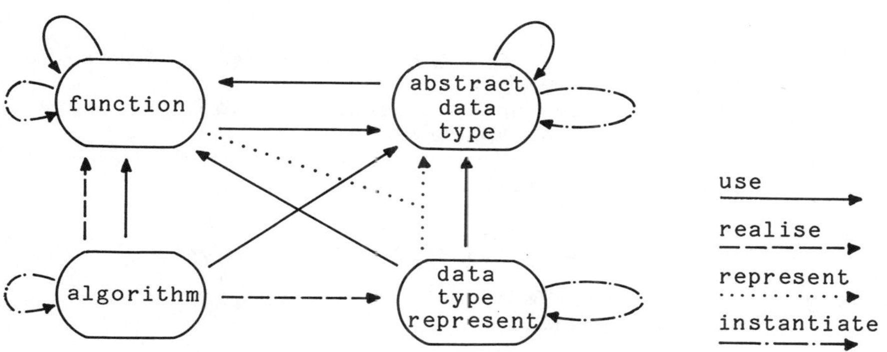

E - Structures and objects for project base evolution
The structures are:
- a project history. The chronological and genealogical
  links among versions are preserved in the PDB as a tree of
  versions.   An object (function, data type, etc.) can
  belong to different versions.   In the project history all
  the defined objects are frozen even those related to
  decisions leading to incorrect developments. The project
  history is in fact a tree whose nodes are labelled with
  development states and whose edges are labelled by development actions.   These development actions, however. are
  not recorded as formal objects ("design decision specification") but only as informal comments. These comments
  are recorded as an attribute of a version.
- an agenda. This structure is used to determine what to do
  next.   According to a "method" what is the set of
  available development actions.
- project management plan. This structure contains information about the project schedule and usage of resources.

<u>9.2.2 Basic relations and primitive actions</u>

    In this section we present the basic means used to
derive a new development state from a previous one. Such
basic actions derive directly from the project base model
itself (objects and basic relations):
- add a new object to an intermediate solution,
- add new property values to an object,
- add a tuple to a relation at an intermediate solution.
- look up in a development state or in the history.

All these basic actions are defined independently of a
method and of the semantic meaning of the objects. Indeed

this semantic control is realised:
- by the use of integrity rules,
- by the use of tools adding automatically these objects or
  the relations depending on the global context and a
  particular action.

There are four kinds of integrity rules, classified
according to two criteria:
- static or dynamic; they apply to states or to transitions
- inherent to the model or optional.

9.2.2.1 Some examples of integrity rules

A - Static and model inherent rules
These rules characterise valid development states.

Example: If an algorithm A realises the function F then A
and F have the same signature.

B - Dynamic and model inherent rules
They deal with transitions between development states.

Example: Data type operators can be realised only by means
of a data type representation.

C - Static and optional rules
These rules are used to restrict the set of valid development states.

Example: An algorithm cannot use more than six functions.

D - Dynamic and optional rules
They are used to restrict the set of valid development
transitions. They reduce the number of activities which can
take place at the same time.

Example: A function must be specified before being realised
by an algorithm.

E - Derived information
Finally the integrity rules can be used to derive information. For example, they can be used to generate the agenda
according to a given "method".
They can also be used to generate explicit tuples of
relations.

Example: If we know the signature of a function, then it is
possible to derive the signature of the algorithm.

9.2.2.2 More global actions: 'design decisions'. Significant changes in the project base -related to a design
decision- are made of successive basic actions which are
supposed to satisfy a given goal. Such global actions could
be seen as "unbreakable" operations allowing the user to go
from a development state to another one.

Example: "Give an algorithmic realisation for function F
with operational properties P"

This global action can be decomposed as follows:
- choice in the project history of the development state
  that we want to develop further,
- computation of the agenda associated with this state,
- choice of a "realisation" description task,
- introduction of the algorithmic description,
- verification of this description with regard to integrity
  rules and requirements and after that generation of a new
  development state.

Moreover giving an object description implies:
- assigning values to properties of that object,
- giving instances of basic relations to link this object to
  others.

Such descriptions and relations can give rise to forward
references to objects that do not yet exist: they have to
be created in a future development state.
Other global actions are needed: for example, representation of an abstract data type, validation of properties,
design plan modification, etc.
An obvious consequence of the action definition possibility
is a need for a means of action composition.

9.2.3 A means of action composition

     The current implementation state of SPRAC is such that
the developer has to manage his development by acting on the
PDB by means of a command language. The only development
control structure is the sequence of commands. The idea of
a development language (4), (7), as an aiding component is
not present in the design of SPRAC. However some basic
tools offered by the system are related to action composition means.

9.2.3.1 Specialised commands. Two commands: "charge" and
"<-" can be used to structure sequential steps of development.
The command "charge" allows the execution of a file of SPRAC
commands.   This possibility can be used to parameterise
steps of development. The level of parameterisation is the
one offered by the host operating system.
The command "<-" allows the developer to store intermediate
results of development step in a variable. These variables
can appear further at the expression level of development
step commands.

9.2.3.2 Definition of integrity rules. One can add
dynamically new integrity rules. We have stated previously
that integrity rules can be used to describe -in a negative
way, by restriction- a method. The integrity rules are also
used by the function computing the agenda. The set of
actions belonging to the agenda can be structured and sized

depending on the method related to the explicit definition
of the integrity rules. The agenda can be seen, in this
perspective, as a function computing, step by step, all the
branches of a design decision tree.

REMARKS:

1) The integrity rules can be used in several ways:

    - to "describe" development methods with dynamic rules
    - to "describe" "good" programs with static rules

2) Depending on the integration of the rules in the system.
   we can have:
    - either a prescriptive model of the software development
      process. The user has to follow what the system
      imposes.
    - or a permissive model. In the later case. the system
      indicates actions to be done.

END OF REMARKS

9.3 CONCLUSIONS

   We have presented a system which is able to assist the
users (what ever they are) in the software development
process. This assistance is:
- restricted from specification to code production
- limited from the point of view of semantics. Indeed the
  Data Base approach chosen for the expression and the
  structuring of design knowledge is obviously weak. The
  Data Base model (objects and relations) is not formalised
  enough; furthermore there is a lack of semantic control on
  the basic actions (adding an object or a tuple). However
  the main ideas of this model are, perhaps, a basis for a
  formal justification of the development of programs with
  the help of automatic or semi-automatic tools. The tuples
  of the relations "realise" or "represent" can be added by
  such automatic tools without the need for a supplementary
  verification step.   In this context a data base model
  could be used to formally document the development steps.
  A step toward this formal description is made possible
  using some results presented in (2) or in formalising the
  input/output relationship of tools such as transformations
  or synthesis systems (4),(1).

   SPRAC is implemented in MacLisp and Prolisp (a dialect
of Prolog in Lisp). It is fully operational for the first
two levels of languages (LF and LA) and for the PDB
supervisor.   The studies now conducted around SPRAC are
twofold:
- integration of the LM language in order to get a complete
  system for experiments about the software development
  process.

- specification of a language of development as suggested in
  2.3. One of the objectives of SPRAC is also to be able to
  be adapted to any specific method of software development.

REFERENCES

1.   P. BERNARD
     "Synthèse d'algorithmes à partir de spécifications
     relationnelles : développement de stratégies"
     Thèse de 3eme cycle. ENSAE. 1985

2.   M. BROY, P. PEPPER. M. WIRSING
     "On relations between programs"
     In Proc. of 4th Int. Symp. on Programming,
     B. Robinet (Ed.), LNCS 83, Springer 1980

3.   J. FOISSEAU et al.
     "Le système SPRAC : expression et gestion de spécifi-
     cations d'algorithmes et de représentations"
     to be published in TSI (Techniques et Sciences
     Informatiques)

4.   J. DARLINGTON
     "The structured description of algorithms derivation"
     in Algorithmic Languages, de Bakker. van Vliet (Eds)
     North Holland, 1981

5.   E. KANT, D.R. BARSTOW
     "The refinement paradigm: the interaction of coding
     and efficiency knowledge in program synthesis"
     IEEE. SE-7, Sept. 81

6.   Z. MANNA, R. WALDINGER
     "A deductive approach to program synthesis"
     ACM. TOPLAS 2 (1), 1980

7.   D. S. WILE,
     "Programs developments: formal explanations of
     Implementations",
     USC-ISI-RR-81-99

8.   G. ZANON, M. LEMAITRE
     "A model for software developement"
     DERI, Internal Report, Nov. 82

Chapter 10

Project support in the Smalltalk-80 integrated environment

L. Peter Deutsch

10.1 INTRODUCTION

The Smalltalk-80^{TM} system is an integrated interactive environment, designed to exploit the power of the bitmapped workstation. The system provides facilities for programming in a powerful object-oriented language, a file system, graphics facilities, a framework for implementing user interfaces, a document editor combining text and bitmap graphics, a mail system, and project support facilities. Formatted text can be transferred freely between any application and any other. The system is designed around a portable virtual machine and has been successfully ported to at least six different processors (three proprietary microprogrammed minicomputers, one mainframe, and two microcomputers), running a variety of different operating systems. For further details on the Smalltalk-80 language and its implementation, see Goldberg and Robson (1) and Krasner (2); for a description of the Smalltalk-80 environment, see Goldberg (3).

Several aspects of the Smalltalk-80 system and language design have influenced the design of the project support facilities currently provided in the Smalltalk-80 system. Consequently, we will discuss these aspects of the larger design before describing the project support facilities in detail. We will also mention some current research in the area of project support facilities within the Smalltalk-80 environment.

"Smalltalk-80" is a trademark of Xerox Corporation.

10.1.1 Philosophy of the Smalltalk-80 System

The Smalltalk-80 system is an ***open system***. There is no distinction between "operating system", "runtime support system", "library", "application", and "user" code. (Many Lisp systems, and the Unix/C environment, are open systems, at least down to some "operating system" level.) The entire system is instantly accessible to the user, in source form, including on-line documentation. Users are encouraged to build applications by adapting or extending existing system or application code: the subclassing mechanism of the Smalltalk-80 language (see below) makes this a powerful means for creating new applications similar to existing ones.

The open system philosophy simplifies project support facilities, because there is no need to support one set of tools for the user and a different one for the operating system or application builder. The testing cycle for tools is shortened, since they can be developed within the environment to which they apply.

The Smalltalk-80 system is ***workspace-based***. (APL and Lisp systems are typically workspace-based.) Rather than reestablishing a working environment by calling up applications and loading files at each session, users normally preserve the entire state of the system from one session to the next. (Such a preserved state is called an ***image*** in our terminology. Unfortunately, the Smalltalk-80 system and documentation use the word "workspace" for a different purpose, not relevant here: see chapters 1 and 2 of reference (3).) External files are currently used for long-term reliable storage of code and other information, but our goal is to extend the workspace metaphor to include the entire body of information accessible to a user, whether stored locally or remotely, and whether shared or private.

The workspace approach has both positive and negative effects on project support tools. On the positive side, a workspace can carry a complete history of its past, so that its user can easily find out (in principle) how the present state of the workspace relates to other workspaces and activities within a project. On the negative side, image files are not designed to be examined by programs, so information can only flow between workspaces through the use of (other) shared external files.

The Smalltalk-80 system has historically been ***single-language***. (APL, Lisp, and BASIC systems are other successful single-language environments, although none of these typically extends the single language down to the level of the I/O drivers and the implementation of such basic concepts as files, tables, or arbitrary-precision integers.) The system is based on a formally specified portable ***virtual machine*** or ***VM***, similar to the Pascal P-System (4): the Smalltalk-80 programming language is implemented by a compiler whose target is this virtual machine.

Project support tools are obviously easier to build if they only need to deal with the semantics of a single language. A topic of current research within our lab is how to evolve the Smalltalk-80 system away from its single-language focus, while sharing as much of the programming environment as possible among languages. We will return to this question in a later section.

The tools described here were developed in ***memory-resident*** implementations of the Smalltalk-80 system, i.e. implementations without any virtual memory facilities. We do not consider this a desirable feature: it is a historical artifact that the large size of our machines' memories (typically 2M bytes) has allowed us to live with.

The main consequence of the memory-resident development environment
is that our project support tools use external files to hold much of their
information. This includes both the log file used to record the history of an
individual workspace, and the data base used to coordinate system evolution. We
will say more about this below.

10.1.2 Highlights of the Smalltalk-80 Language

The Smalltalk-80 language follows the ***object-oriented*** programming
paradigm, derived from Simula-67 (5) and related to Actors (6). Procedures and
data structures in conventional programming languages are usually defined in
isolation from each other, or coupled together by some third mechanism such as
"modules" or "abstract data types". In a Smalltalk-80 program, on the other
hand, each type of object is defined by a ***class*** which specifies the names of its
state variables. The objects defined by a class are called the ***instances*** of that class,
and the state variables are called ***instance variables***. Every procedure (or ***method***,
in Smalltalk-80 terminology) is associated with a class. Only the methods of a
class can directly access the state of an object of that class: external access is only
possible by calling a method. This constitutes a powerful encapsulation
mechanism.

In the Smalltalk-80 language, procedure names are left in symbolic form
until invocation time, and interpreted according to the class of the (first)
argument. For example, in the expression `3 + 4`, the procedure name "+" is
looked up in the method dictionary of class `SmallInteger`; in the expression `3.0
+ 4.0`, "+" is looked up in class `Float`. We use the terminology of Actors, in
which the procedure name is called a ***message name*** (or sometimes simply a
message) and the argument whose class is used to look up the name is called the
receiver of the message. However, unlike Actors, our "message sending"
operation is synchronous, and should really be considered a form of polymorphic
procedure call.

A Smalltalk-80 class can be defined as a ***subclass*** of an existing class. If C is
a subclass of D, then C may add instance variables to D, and may add or redefine
methods of D. If a message is sent to a receiver which is an instance of C, but C
does not define the message, then the message is looked up in D, and so on. Most
commonly, C defines some operation left abstract in D. For example, `Integer` is a
subclass of `Number`; `SmallInteger` is a subclass of `Integer`. Another example,
drawn from the Smalltalk-80 user interface, is `CodeController` (the editor for
program text) which is a subclass of `ParagraphEditor`, adding some operations
that are only meaningful when editing programs. This illustrates another use of
subclassing, namely to specialize or modify an existing class for a new application.

10.1.3 Highlights of the Smalltalk-80 Programming Environment

The Smalltalk-80 user interface uses the now-familiar style of overlapping windows, pop-up menus, and a bitmapped display terminal, controlled primarily with the "mouse" pointing device. (In fact, our laboratory was largely responsible for developing this style.) As in most such interfaces, windows are *function-based*. In other words, to take information being viewed through one set of functionality (e.g. a piece of source code) and apply another kind of function to it (e.g. send it in a mail message), it is necessary to manually "copy" or "cut" the information from its source, and "paste" it into another window. In contrast, in the pioneering NLS interface (7), windows are *document-based*: the content stays where it is, and any application can be "wrapped around" or used to view any document. We intend to experiment with this different approach to function integration in the future.

The programmer's primary interface for reading and writing code is a window called a *browser*. This window presents a four-level tree structure, with each level represented by a menu. The top level is a list of categories, in which classes are grouped together by convention (e.g. there is a category called "Graphics-Primitives" that defines such basic graphical classes as `Point` and `Rectangle`.) The next level indicates the classes that are members of the category selected at the first level. The third level groups methods within a class, again into categories chosen by convention. The final level lists methods within a category. The method selected at this level is displayed in a separate window for editing.

Since the browser is the only mechanism by which a programmer normally writes code, it implements some of the project support facilities, such as logging actions related to program creation and modification. It also provides the interface for special queries such as "What methods send this message?", "What methods implement this message?", and "What methods in this subtree of the class hierarchy reference this instance variable?"

Limited space prevents us from discussing the Smalltalk-80 debugger and the other tools in the programming environment. For a full description of the Smalltalk-80 user interface, see reference (3). The user interface for the project support facilities is described in chapters 22 and 23 of (3); the system evolution (version management) facility is described in chapter 15 of (2).

10.2 EXISTING FACILITIES

The workspace-based nature of the Smalltalk-80 environment leads us to divide its project support facilities into *individual* (primarily for use by a single user within a workspace) and *group* (primarily for shared use by multiple users, based on external files.) We discuss the specific facilities of each kind in detail below.

The workspace-based environment makes individual project support facilities particularly easy to implement, but does not help with group facilities. Individual facilities are based largely on the manipulation of Smalltalk-80 objects within a workspace. These objects represent things like classes, methods, and user actions. The full power of the Smalltalk-80 programming language and system classes is available to implement their functionality. In contrast, group facilities are based largely on external files. Much of the implementation of these facilities is concerned with converting objects between external (text file) and internal (object) representation, with synchronizing access to external files, and with implementing the client side of already-defined interfaces to external services (such as the mail system and file servers). This large body of code contributes little to the functionality of the objects being manipulated.

10.2.1 Individual Project Support Facilities

10.2.1.1 Historical record.

The principal project support tool for individuals is a historical record of every "significant" action taken since the creation of a workspace. The system keeps this record in two forms: a summary, kept within the workspace, of discrepancies between the contents of the workspace and the external files used for permanent storage, and a complete log, kept on an external file, of actions in historical order.

"Significant" actions include defining, redefining, or removing classes or methods, or the on-line documentation that accompanies them, or executing any expression through the user interface. The system notices these actions in a variety of places, some of them at the user interface level, some of them deeper down in the implementation: we would like to replace this situation with a more consistent design, perhaps modelled on the history facilities of Interlisp (8).

The internal summary of significant actions is held in an object called a `ChangeSet`. This object is basically a set of pairs of the form <object affected, type of action>. "Object affected" is a class or a <class, message> pair; "type of action" may be creation, modification, or removal. Mutually cancelling actions (for example, the creation and subsequent removal of a method) are detected and removed from the `ChangeSet`. The purpose of the `ChangeSet` is to provide the user with enough information to be able to write changed classes (or incremental changes) back onto external files. When the user does this, the corresponding entries are removed from the `ChangeSet`.

The external log is simply a text file. Each action is recorded in the form of an expression in the Smalltalk-80 language: evaluating the expression has the same effect as the original action. Logging an action consists of appending the text of the expression to the log file, and then forcing any buffered pages out to the storage medium (local disk or file server). This makes the log extremely robust. In fact, the log serves three separate purposes:

- During normal operation of the system, the source code for methods is not held in the workspace, but is read in from the log file when the user examines it. (This is an expedient to circumvent the lack of virtual memory facilities in our Smalltalk-80 implementations.) The workspace holds information giving the log file position for the source code for each method.

- If the system crashes, it is possible to restore the state of the workspace from the last snapshot (see below), and then replay the log since the point corresponding to the time at which the snapshot was made. The user can replay the log either in toto or selectively, using the `ChangeList` tool described below.

- Since <u>all</u> versions of the source code for a method or class appear on the log file, it is possible to examine and even re-install older versions. (The system currently provides only a rudimentary tool for examining old versions of a method.)

The primary tool for examining and replaying the log is an object called a `ChangeList`, which is simply the result of parsing a segment of the log and constructing a sequence of objects to represent the actions. The user interface to the `ChangeList` allows the user to selectively view the actions relating to a given class, message, and/or action type (create/modify/remove). The user may edit the list by removing selected actions, and may replay individual actions or the entire (remaining) list. A single command to the system creates a `ChangeList` containing all the actions since the last snapshot and opens an interface to it on the screen, which is exactly what the user wants when restarting after a crash.

10.2.1.2 <u>Projects.</u> The Smalltalk-80 system provides some support for a user working on several different projects in a single workspace. At any time, the user can create a new `Project` object and a corresponding icon on the screen. The only command for a project is to "enter" it. This brings up a separate screen, which the user may now format in any desired way. The system command "exit project" returns the user to the former project. In this way, projects within a workspace form a tree: "entering" a project descends one level in the tree, while "exit project" ascends a level.

In addition to separate screens, each project has its own `ChangeSet`. This allows the user to separate to some degree the actions taken in different projects. (Note that actions taken in any project affect the same underlying information, e.g. a class created in one project is visible in all projects.)

The multiple project facility is not used heavily. There are several possible reasons for this.

- Lack of virtual memory means there is usually not enough room for more than one substantial project underway in a workspace.

- Since all projects share a single set of classes and methods, it is inconvenient to work on several projects that affect any of the same classes.

- Switching projects requires conscious effort because they have completely separate screens. For example, one cannot have two browsers on the screen that register their changes in different projects.

- If one accidentally makes a change in the wrong project, there is no support for moving it to another one.

- The action log does not record the project under which each action was taken, so assigning actions to projects when replaying the log must be done manually.

10.2.2 Group Project Support Facilities

<u>10.2.2.1 Files.</u> One way for a group working on a project within the Smalltalk-80 environment to share information is to simply share or copy the stored representation of a common workspace. A workspace is represented as a pair of files: the image file, containing a copy of the contents of memory, and the action log file described above. There is a command that creates an image file from the current contents of memory: such a file is sometimes called a "snapshot". Image/log file pairs are customarily copied between workstations and file servers using an external operating system.

We have found that sharing a workspace is a good strategy when the group working on a project is very small, and when the group members are not all working full-time on the project. Otherwise, contention for the use of the workspace becomes a bottleneck.

The other, and more usual, method of sharing information is via files that contain source code. One file normally contains either a group of related classes, a single class, or a collection of incremental changes to be executed in the context of some workspace. (The programming environment supports the creation of all three kind of files.) Such files are time-stamped with the system version and the date and time they were created. However, since there is currently no support in the Smalltalk-80 system for keeping track of the contents of a workspace (other than in terms of the individual actions that have been carried out), there is no way of finding out the prerequisites of a given piece of source code, in terms of what classes or changes it assumes are present in the workspace. This is a significant weak point in our present programming tools.

<u>10.2.2.2 System evolution.</u> The Smalltalk-80 system supports the orderly creation of a single, linear sequence of system versions or releases. A shared data base, represented by a set of text files stored on a file server, records bug reports, bug fixes, and proposed system enhancements ("goodies") submitted by users using tools present in the standard system. The data base also includes a record of which changes were incorporated in producing system version N + 1 from version N.

The version management system is designed to have multiple instances. Each instance has its own set of external data base files. For example, the group that supports the Smalltalk-80 system as a product has a different data base, and a different sequence of system versions, from the research center; within the research center, there is one instance of the version manager for the Smalltalk-80 system proper, and another one for the experimental cross-programming tools.

The first steps in producing a new version are partly social and partly mechanical. In our laboratory, a "release czar" chooses the interval of releases (typically one to three months), and proposes (via electronic mail) which bug fixes and goodies should be included in the release. A meeting of all interested lab members produces the final list. Typically, much of the work of choosing the list is done via electronic mail in the week preceding the release meeting, and the meeting itself resolves differences of opinion in only an hour or two.

Once the set of changes that will go into a new release has been chosen, the release czar uses a specialized command in the ChangeList to examine the set of changes and detect any conflicts (several bug fixes or goodies may affect the same method, or one may modify a method that another removes.) Conflicts may require minor manual editing of the source files. Finally, a command to the version manager incorporates the changes into the workspace and registers the creation of a new version in the data base. It takes the release czar about a week of work to accomplish these tasks.

Over the year and a half in which the release tools have been available, we have produced 9 releases of the system. Each release has added an average of 44K bytes of source code and 28K bytes of object code to the system. (The system might well have grown faster if the lack of virtual memory in our implementations did not impose a hard limit on the system's maximum size.) The total size of the system is approximately 2.5M bytes of source code and 1M bytes of object code. About 25 people use the system actively within the laboratory.

The rapid evolution of the system, combined with the lack of tools for understanding the prerequisites or system effects of a given piece of code, leads to a situation where any piece of code not incorporated into the system itself tends to "decay". Most code over a year or two old will not run in a current system. This is a real problem, although several factors mitigate it in our work environment:

- The debugging and analysis facilities within a single workspace make it relatively easy to fix the problems in old code when they arise.

- We keep all system versions available, either on-line or on archival storage, indefinitely. This allows one to run old applications in their original system if necessary.

- Almost every application or function of substantial utility eventually gets incorporated into the system. This is feasible because we are a relatively small group and do not produce new features at a rate that overwhelms the ability of the system (and us as users!) to absorb them.

<u>10.2.2.3 Integrated mail system.</u> While not a project support tool per se, the electronic mail facilities present in the Smalltalk-80 environment assist in project support by reducing the need for other, more specialized tools. For example, every few days the system administrator runs a program that uses the mail system to distribute a summary of recent bug reports and bug fixes to all mailboxes on the "Smalltalk80Users" distribution list.

<u>10.3 ONGOING RESEARCH</u>

<u>10.3.1 Replacing External Files</u>

The use of external files to store the data base for group project facilities is a major conceptual bottleneck in the design and implementation of our programming tools. Ideally, objects shared between users would have semantics and performance similar to objects local to a single user. This has led us to the notion of some kind of shared, permanent, object-oriented memory to replace our current environment of mail and file servers. We are currently trying to define what properties this memory should have in order to support maximally transparent sharing of information between users. Such properties include the desirable properties of existing file and data base servers (synchronization, robustness, rapid retrieval from aggregates, etc.) as well as making shared objects behave as much as possible like local ones.

<u>10.3.2 Retargeting the Programming Environment</u>

As mentioned earlier, we are experimenting with retargeting the Smalltalk-80 programming environment to work with multiple languages. Our first steps have involved source languages whose semantics can be represented easily within the Smalltalk-80 VM. Such languages compile into the Smalltalk-80 language as an intermediate step. Two examples currently under investigation are a compiler-writing language and a finite state machine or production system language.

We are also exploring the use of the Smalltalk-80 programming environment for languages with different target semantics, such as assembly languages or conventional compiler languages like Pascal or C. Even in this case, the tools described here still apply to some degree, since these tools deal almost entirely with source code and do not interpret its semantics in detail.

10.3.3 Supporting the Group Process

In addition to our work on the Smalltalk-80 programming environment per se, we are pursuing a design methodology research project that uses the Smalltalk-80 environment simply as an implementation vehicle. The purpose of this project is to explore using electronic media to serve organizations that carry out design projects. We believe that electronic media can be used to record most of the significant events of a design project, that the records can be used as the primary unit of communication within the project, and that use of the records will enable fundamental transformation of the process by which projects are carried out.

This design project is necessarily long-term in nature. The introduction of electronic media into the fabric of organizational process is linked with introduction of both electronic infrastructures (such as computing networks and video) and participatory and self-organizing social practices. The introduction of these changes takes years, and our research is about bringing these long-term changes under conscious control.

10.3.4 Remote Laboratory Experiment

We are establishing a second branch of our laboratory in a city over 1,000 km away. The purpose of this experiment is to explore computer-based tools that can fulfill some of the functions of the informal, personal communication that now exists within our rather close-knit group. This will involve a greater focus on electronic communication integrated with computer facilities, as opposed to the facilities provided within individual workstations.

REFERENCES

1. Goldberg, A., and Robson, D., 1983, 'Smalltalk-80: The Language and its Implementation', Addison-Wesley Publishing Co., Reading, MA, U.S.A.

2. Krasner, G., ed., 1983, 'Smalltalk-80: Bits of History, Words of Advice', Addison-Wesley Publishing Co., Reading, MA, U.S.A.

3. Goldberg, A., 1984, 'Smalltalk-80: The Interactive Programming Environment', Addison-Wesley Publishing Co., Reading, MA, U.S.A.

4.   Ammann, U., et al, 1975, 'The Pascal (P) Compiler Implementation Notes',
        Institut Fur Informatik, Eidgenossische Technische Hochschule, Zurich,
        Switzerland.

5.   Dahl, O.-J., Myhrhaug, B., Nygaard, K., 1970, 'SIMULA--Common Base
        Language', Norwegian Computing Center, Oslo, Norway.

6.   Yonezawa, A., Hewitt, C., 1977, 'Modelling Distributed Systems', M.I.T.
        Artificial Intelligence Laboratory publication # 041428, Cambridge, MA,
        U.S.A.

7.   Watson, R. W., 1976, 'User interface design issues for a large interactive
        system', AFIPS Conference Proceedings, Volume 45, pp. 357-364, AFIPS
        Press, Montvale, NJ, U.S.A.

8.   Masinter, L. M., ed., 1983, 'Interlisp Reference Manual', Xerox Special
        Information Systems, Pasadena, CA, U.S.A.

Chapter 11

Abstract datatypes and the IPSE database

Adrian R. Jackson

11.1   INTRODUCTION

As part of the U.S. Department of Defense's (DoD)
programme to standardise on a common high order language for
progamming embedded systems, the need for both the language
itself, Ada[1] (Ichbiah et al.(1)) and an environment to
support the production of Ada programs was recognised. A
specification for such an Ada Programming Support
Environment (APSE) has been produced by the DoD and is known
as STONEMAN (U.S. DoD (2)).

STONEMAN contains a description of the requirements for
all aspects of an APSE including those for an APSE database.
It is demonstrated here that apart from certain language
specific elements, these APSE database requirements form the
basis of a more general set of IPSE database requirements.

Using these requirements an IPSE database structure has
been designed, which relies heavily on the use of abstract
data types to provide access control and protection
facilities.

11.2   REQUIREMENTS FOR AN IPSE DATABASE

11.2.1 The STONEMAN Requirements

The STONEMAN document presents the APSE specification
on three levels, those of the full APSE, the Minimal APSE
(MAPSE) and the Kernal APSE (KAPSE). The MAPSE is the
database plus the minimal set of tools required to perform
the programming support function. The KAPSE level of the
specification defines the database, and the interface
requirements to support the use of tools on the database
contents. Data in the KAPSE database is kept in a container
known as an object, which may represent any of the varieties
of data to be stored e.g. program text, documentation,
executable programs etc.

The KAPSE database requirements specified in STONEMAN
are enumerated below.

(a) Every object in the database must have a unique name.

---

[1] Ada is a registered trademark of the US Government, AJPO

(b)   Objects must be capable of being grouped together as version groups, and one member of such a group may be designated as the preferred member.

(c)   There will be no format restrictions on the content of an object.

(d)   The database shall permit relationships to be maintained between objects.

(e)   Objects in the database shall have attributes, the following of which are mandatory,

   – a history attribute which records the manner in which an object was produced, and provides the necessary information for a system of configuration control,

   – a categorisation attribute defining the likely usage of the information in the object,

   – an access rights attribute to be used to determine potential users access capabilities.

(f)   Some provision must be made for archiving and retrieval of parts of the database.

(g)   Reading and writing of objects must be performed by the standard Ada INPUT-OUTPUT package.

(h)   Objects may be collected together to form configuration groups which are themselves objects.

(i)   The information in the database may be partitioned into related collections.

(j)   The database shall support source and part compiled Ada libraries.

(k)   The consistency of the database must be preserved.

Considering the above list, the only items which are directly related to the Ada language are (g) and (j). Item (g) concerns the input output facilities of Ada and states that the information in the database must be accessible using the standard input output facilities of Ada. Item (j) states that the database must contain facilities for the storage and manipulation of Ada libraries. Neither of these requirements need be satisfied by the database itself. In the case of (g), if the package INPUT-OUTPUT is implemented to suit the database rather than vice versa then a general purpose set of access mechanisms provided by the database should be sufficient. In the case of (j), this may be satisfied by requirement (c), which implies that Ada libraries may be stored, and the provision of suitable tools for accessing and manipulating these libraries. Taking

these two considerations into account, it can be seen that
the overall set of KAPSE database requirements is not
necessarily language specific, and that these requirements
could apply to any programming support environment provided
that there are similar motivations concerning reliability
and control of the database to those of the DoD.

11.2.2 The STONEMAN Requirements as IPSE Database Requirements

Examination of the above requirements also shows that
they are not particularly related to the task of programming
alone. In particular items (g) and (j) again are the only
ones directly related to programming. The rest of the
requirements could apply equally to any form of information
or collection of information in the database e.g.
documentation, programs, configurations of software etc.
Requirements (b),(d),(e),(h) and (i) specify various
means of collecting information together and specifying
relationships between objects or collections of objects.
These structuring and connecting facilities may be applied
to data at any stage in a project's lifecycle and may also
be used to relate the various stages to each other. For
example source modules implementing the same specification
in different ways, may be related by partitioning (i.e. by
being in the same collection of objects) to the
specification of the function of the modules. Each separate
source module could be related to its design documentation,
either by partitioning or the use of the history attribute
(i.e. an indication of which design a source is derived
from). The designs may be related to the specification in a
similar manner.
Figure 11.1 illustrates the situation, partition "A" is
used to group the sources and their specification together
and partition "C" groups the designs together. The sources
are related to their designs either by their history
attributes or by partitioning as shown by partition "B".
Partitioning may also be used to form version groups of
objects, in this case the use of attributes is necessary to
impose some ordering between the members of a version group.
These general structuring facilities and the fact that
STONEMAN deliberately addresses all stages of the project
lifecycle indicate that the APSE would be better
characterised as a project support environment specific to
Ada rather than a programming support environment. Having
established that the only Ada specific KAPSE database
requirements are those of I-O support and library
management, then if these functions are performed by special
purpose tools and the provision of appropriate Ada packages,
the KAPSE database fulfils the requirements for the kernel
of an IPSE database.
Several APSE designs have been produced which are
highly Ada specific, most notably from Intermetrics (3) and
the UK Ada Study Group (4). The proposal from Intermetrics

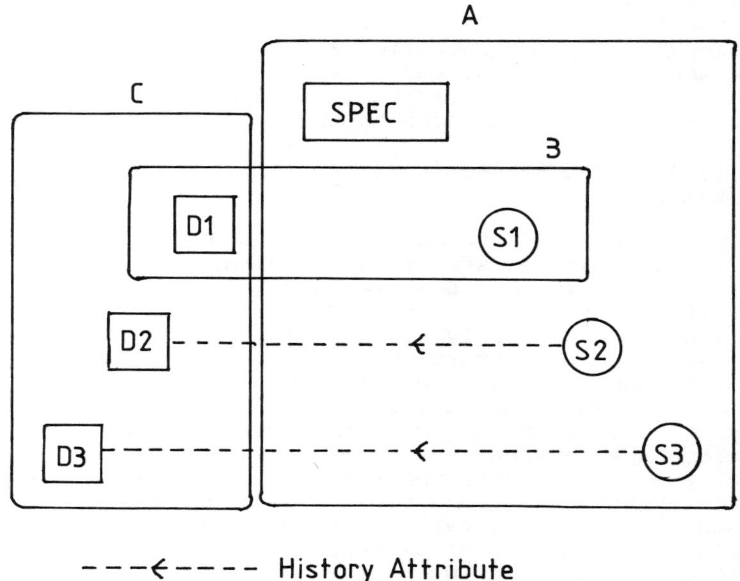

Fig. 11.1 Structuring information in the database.

satisfies the KAPSE database requirements, but makes great
use of an abstract syntax tree representation for Ada
programs and the representation of programs in this manner
is intimately connected with the overall database structure.
The UK Ada Study Group APSE design includes as part of the
database a model for the representation of Ada libraries.
In addition an examination of, and proposals for the
satisfaction of the lifecycle support requirements of an
APSE has been published (McDermid and Ripken (5)). It is
notable that the major language specific areas of this study
are those at the module design and programming level.
        It is my contention that a well designed APSE database
and a well designed IPSE database are equivalent, with the
respective environments differing at the tool level i.e. in
the support provided by tools for different methods and
languages.

11.2.3 Proposed Requirements for an IPSE Database

        The following minimal requirements for an IPSE database
are proposed.

(a)   Basic structuring facilities i.e. some means of
       grouping and relating information into structures such
       as version groups, configurations, partitions etc.

(b)   A means of accessing the data contained within the
        above structures i.e. some means of identifying and
        using items of information at any level of structuring
        imposed by the database.

(c)   A means of controlling access to the information in the
        database, both according to who (or what) uses it and
        how it is used. This control to be provided at any
        level of information structuring within the database
        deemed necessary by particular project requirements.

(d)   A means of storing information about the data in the
        database (attributes) and a means of using this data.

(e)   A mechanism by which tools may be provided to operate
        on the data in the database, in order to provide the
        facilities and operations of the IPSE. This mechanism
        is to be open ended, in that new tools may be added to
        the environment when necessary.

(f)   Some means of ensuring the security and consistency of
        the database.

11.3   THE ABSTRACT DATA TYPE AS A PROTECTION AND ACCESS CONTROL MECHANISM

11.3.1 Basic Principles

An abstract data type (Jones (6)) may be defined as a
set of possible values that the type may possess, together
with the operations defined to apply on those values. The
operations alone completely characterise the type. A design
for an IPSE database is presented here which uses
collections of tools and the operations represented by them
to define abstract data types, which are used to provide
access control and protection facilities for the information
in the database.

Some means must be provided within the database of
defining operations on particular types and of grouping
these operations together. In addition a mechanism must be
provided for specifying particular instances of items of a
type to be operated on.

These facilities are provided in Ada by the package
construct, which enables a concrete representation of a type
and the operations on that representation to be defined.
The actual representation of the type is hidden from
potential users, whose sole interface to the package is
through the operations provided. A similar method is used
in this database to restrict access to data items.

11.3.2 The Database Structure

Before an access control mechanism can be defined the
fundamental facilities for structuring and representing data
in the system must be defined.

Any container of information within the database is
known as an object, which consists of two parts, its basic
information content (the body) and a set of attributes
describing the object. Objects are of two types, simple and
compound. Simple objects are the basic containers of
information and the body of a simple object contains a
simple string of bytes. Compound objects are the basic
structuring facilities and the bodies of these contain
references to further objects, either compound or simple.
The data in the system is organised in the form of a
directed graph. Each node of the graph is an object.
Terminal nodes, with no arcs of the graph leading from them
are represented as simple objects. Non-terminal nodes are
compound objects. The out arcs from a compound object lead
to the objects said to be members of it (those referenced
within its body). All the arcs within the database graph
are labelled and one compound object is distinguished as the
root object, from which any other object in the graph may be
reached by following a sequence of arcs. The root is also
defined not to be a member of any other compound object.
Each object in the database possesses a set of
attributes. Attributes are represented as a set of name,
value pairs where the value is a string of bytes.
Attributes may be used either to hold information about an
object as in the case of a categorisation attribute, or
additionally to express some relation to other objects, as
in the case of a history attribute which records the
derivation of an object from other objects.
Figure 11.2 represents the same structure as figure
11.1 with the exception of the relationships formed by
history attributes. This structure is a small part of the
database representing a small part of a project. The
ellipses represent compound objects and the squares and
circles represent simple objects. The compound objects are
used to provide a mechanism for grouping objects together
into configurations, version groups, partitions etc. This
is akin to the use of directories in conventional filing
systems, with the major exception that an object is
permitted to belong to more than one compound object.
The fact that an object may belong to more than one
compound object permits different views of areas of the
database to be obtained. For example in figure 11.1 one
source appears in both partition "B" and partition "A" and
one design appears in both partition "B" and partition "C".
This is represented in the graph of figure 11.2 by using
compound objects to form the partitions. Thus the source
object identified by the arc "S1" from partition "A" is also
identified by the arc "S" from partition "B". Therefore the
source object may be viewed as either a member of a
collection of related sources, or as a member of a partition
which associates it with its corresponding design. These
facilities enable multiple views of data structures to be
obtained without duplication of the underlying data.
The other main use of compound objects is to group
objects together as instances of particular data types,

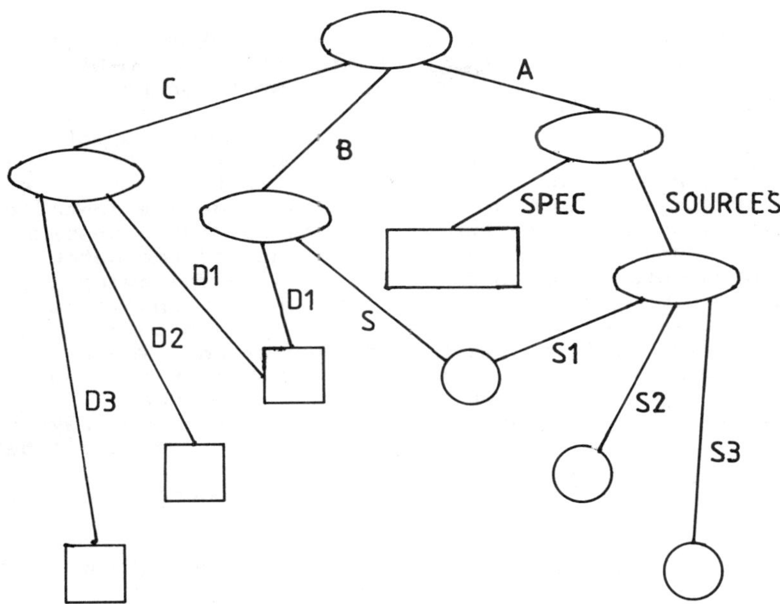

Fig. 11.2 An Example of the Database Structure.

providing regions of the database upon which specified
collections of tools may operate.

The tools which operate on the objects of the database
are themselves present as objects, this enables them to be
grouped together as compound objects which define abstract
data types i.e. the operations provided by the groups of
tools define the type. These compound objects are known as
"toolsets".

11.3.3 Object Naming Conventions

Any object may be uniquely identified by a sequence of
arcs which when followed from the root arrive at that
object. The sequence of labels which identify these arcs is
known as a pathname, using UNIX[1] terminology. For example
in figure 11.2 if the top compound object is assumed to
depend from the root by an arc labelled "Project1", then the
source module shared between partitions "A" and "B" may be
identified by the pathname "Project1 A SOURCES S1" or
alternatively by "Project1 B S". An object is uniquely
specified by a pathname in the sense that each pathname
specifies only one object. This is not to say that an
object may not be specified by more than one pathname.

---

[1] UNIX is a trademark of Bell Laboratories.

11.3.4.The Role Mechanism

Associated with each user of the database is a set of "roles", which are used to determine which areas of the database a user may reference and in what manner. Each role represents a particular toolset and a list of compound objects. At any given time a user may choose to act in one role using the tools in the toolset on one of the database areas defined by the compound objects in the list. The compound objects in the list effectively define instances of data upon which the tools are constrained to operate. Each compound object defines a sub-graph of the database, the components of which may be reached by following arcs from the compound object. Furthermore each individual tool in the toolset may be constrained to operate only on objects at specified positions within the sub-graph. These specified positions are kept as attributes of the tool in question. The tools also perform checks on the categorisation attributes of objects in order to prevent accidental misuse e.g. editing an object containing binary data.

The full pathname of an object referenced by a tool is formed by concatenating three components,

(a)   The pathname of a compound object associated with the tool in a particular role,

(b)   A sequence of arc labels held as an attribute of the tool, which will select a particular area of the subgraph defined by (a),

(c)   A sequence of arc labels provided by the user to specify and object within (b).

A user will select a particular role by asking the system to change the user's current role to one described by the name of a role and the compound object below which the objects to be manipulated reside. If this particular combination of role and compound object do not exist for this user, permission will be denied.

Thus a user's access to the database is completely determined by his current role, and the resolution of the access control mechanism is determined by the nature of the tools provided to the role, and the level at which the tools access the database i.e. whether the objects referenced by the tools themselves define large or small areas of the database.

11.3.5 An Example

Using the area of the database illustrated in figure 11.2 and assuming that the top compound object depends from the root by an arc labelled "Project1" and also assuming that two users have access to this sub-graph.

"User A" who has the roles of Manager, Checker and

Specifier and "User B" who has the roles Programmer and
Designer. The following simple tools are available.

     BROWSER - for read only access to objects containing
            text
     EDITOR - to provide a means of altering the content of
            text objects
     REVISER - to produce new versions of objects
     OBJECT MANAGER - for performing general administrative
            functions
     COMPILER
     LIBRARY MANAGER
     DEBUGGER

The following toolsets for each of the above roles may
be defined

ROLE	TOOLS
Designer	BROWSER for inspecting the specification, EDITOR and REVISER for creating and amending designs.
Programmer	BROWSER for reading designs and COMPILER, EDITOR, REVISERS, DEBUGGER for producing and testing new programs.
Manager	OBJECT MANAGER for administration.
Checker	BROWSER for viewing the contents of objects
Specifier	EDITOR and REVISER for amending and creating new specifications.

All of these roles will refer to the compound object
"Project1".

Taking "User B" as an example the tools belonging to
the role Programmer will have the following partial
pathnames stored as attributes, the BROWSER will have "C"
and COMPILER etc. will have "A SOURCES". This means that
all attempts to access objects using the COMPILER will
automatically have the pathname "Project1 A SOURCES"
prefixed to any names supplied by the user, thus restricting
this tool to apply to the required sources. Additionally
the BROWSER will provide read only access to the designs in
the partition "Project1 C".

Alternatively all the tools of this role may use the
partial pathname "B", which would restrict access to the
design "Project1 B D1" and the source "Project1 B S".
However this would permit access by the EDITOR to the
design, to prevent this the BROWSER could use the partial
pathname "B D1" and the EDITOR etc. could use the partial
pathname "B S", in which case any use of the EDITOR would be
constrained to the single object "Project1 B S". Any
sequence of labels supplied by the user, must, when appended

to the pathname generated by the role and tool, form a valid
pathname, in order to specify an object.
     In the case of "User A" in the role of Manager, he will
require access to all objects belonging to the project and
the tool OBJECT MANAGER would use no additional partial
pathnames. "User A" would then be able to refer to any
object in the project by supplying an appropriate sequence
of names.
     Tools which refer to more than one object may have
partial pathnames specified for each object. e.g. a compiler
may have the locations of its sources, object and library
determined independently.

11.3.6 Summary

     The nature of the access permitted to a user of the
database is completely determined by his current role and
the object that role is operating on i.e. by an abstract
data type and a particular instance of data. The resolution
of the access control mechanism is controlled by the nature
of the tools provided to roles and the level at which the
roles access the database. The system described is intended
to be the kernel of an IPSE database and as such is
independent of any method or lifecycle models and requires
the design and installation within the database of suitable
tools to support whatever methods are chosen. However the
abstract data type mechanism described is intended to
parallel the package and generic package facilities offered
by Ada.

11.4   CONCLUSION

     The system as described is currently being implemented
using UNIX (Ritchie and Thompson (7)) as a base. In
addition to the access control facilities, provisions are
also made for handling multiple versions of objects and for
recording derivation histories of objects.

REFERENCES

1.   Ichbiah, J., et al., 1983, 'Reference Manual for the
      Ada Programming Language', U.S. Department of Defense.

2.   U.S. Department of Defense, 1981, 'STONEMAN,
      Requirements for Ada Programming Support Environments.

3.   Intermetrics Inc., 1981, 'Draft Development
      Specification for Ada Integrated Environment'.

4.   Department of Industry, 1981, 'UK Ada Study Final
      Technical Report', Vol.2.

5.   McDermid, J., and Ripken, K., 1984, 'Life Cycle Support
      in the Ada Environment', Cambridge University Press,
      Cambridge, England.

6.   Jones, C.B., 1983, 'Software Development – A Rigorous
      Approach', Prentice Hall International, London,
      England.

7.   Ritchie, D.M., and Thompson, K., 1978, 'The UNIX Time
      Sharing System', Bell Systems Tech.J., 57(6), 1905–29.

Chapter 12

Rapid prototyping using an abstract data store

J.W. Hughes and M.S. Powell

12.1 INTRODUCTION

The technique of stepwise refinement or *"divide and rule"* is a widely accepted strategy, not only for the design of software systems, but also for the derivation of requirements specifications for computer systems. Conventionally, refinement of a specification takes place in conjunction with an interactive dialogue between the systems analyst and the user. Prototyping enhances this process by demonstrating graphically, to the user, what the proposed system will look like. It is only a viable tool in the dialogue, if the prototype can be readily modified to reflect the refinement process. This paper describes how a highly interactive interface to an Abstract Data Store (ADS) can be used for such a flexible prototyping facility.

The ADS is a software tool which supports abstract data types together with flexible mechanisms for specifying, for each abstract data type, alternative concrete representations appropriate to the different physical media. These mechanisms facilitate the definition of types, the specification of their alternative representations and the creation and manipulation of their values in a persistent fashion, together with the generation of chosen representations on the appropriate media as required. The media supported may include such things as disks and visual displays and collections of these connected together via some form of network.

The user's interface to the ADS is structured so that, at any time, he sees an object together with a menu of just those operations he can sensibly perform on it, dependent on what kind of object he is looking at. The means of activating the operations is uniform, thus simplifying what the user has to learn in order to become adept. It relies heavily on selection using a cursor, which minimises the number of key strokes for non experts and gives an encouraging sense of "touching" the objects manipulated.

The ADS lends itself to data driven software design methods which rely on modelling the environment, which software is to manipulate, in terms of abstract data types. The level of detail modelled may be that of a prototype initially, but is readily extended by further refinement to that necessary for a full implementation. When such systems are implemented conventionally, suites of programs are required to convert between human readable or backing store/filing system compatible representations of the data types they manipulate and internal machine representations of these values. Using the ADS this process is provided automatically by the viewing mechanism. Thus, using the ADS, activities such as data capture and report generation become trivial and rapid prototyping is a natural component of system design.

In summary, the ADS brings together and unifies three separate developments in the state of the art of software development aids,

i) Although modern languages permit strong, user defined, typing of objects, the objects exist only during program execution, whereas they are often required to persist between and independently of program executions. PS–Algol[1, 2] goes some way towards this requirement but does not support the rich set of type constructors introduced in Pascal like languages.

ii) The concept of abstract type, as developed by Hoare[3] and incorporated in programming languages by others[8, 9] still reflects the known properties of immediate access store, whilst more recently, the development of data base technology has resulted in a completely different vocabulary for expressing the same concepts when applied to more long–lived data values. There is obviously a need to unify the terminology of data bases on the one hand and programming languages on the other. In doing so it should be possible to free abstract types from the biases imposed by their medium of representation.

iii) Despite, or because of, the fact that modern languages allow the level of data abstraction to be raised by user defined types, vast amounts of effort go into writing data validation and report generation programs whose sole purpose is to convert between differing concrete representations of values of some abstract data type. Although data validation and report generator tools are available for languages with fixed type structures they have not kept pace with data abstraction mechanisms. Once abstract types are freed from medium imposed biases, then it becomes not only useful, but also viable to let the user define concrete representations of abstract data types, appropriate to each medium.

The rest of the paper explains how the ADS integrates the desirable facilities indicated above, by first elaborating on the distinction between abstract and concrete types, then describing the specification of concrete representations. These sections are followed by a detailed example showing how the ADS has been used for prototyping, which also illustrates the user interface.

12.2 ABSTRACT AND CONCRETE TYPES IN THE ADS

The data types supported are modelled on those expounded by Hoare[3]. The simplest types comprise atomic values, conventionally known as scalars. Composite types are constructed from simpler ones by forming:-

i) Cartesian products of types, whose individual values are ordered n–tuples of component values, conventionally known as records.

ii) Functions from type to type, whose individual function elements allow selection of the image associated with each domain value, conventionally known as arrays when the functions are total and mappings when they are partial.

iii)   The star closure of a type, whose individual values are sequences of values of the component type, conventionally known as lists.

iv)   The power set of a type, whose individual values are sets of values of the component type, conventionally known as sets.

v)   The discriminated union of types, whose individual values are selected from any type in the union. These are conventionally known as variant or union types.

vi)   Recursive definitions allow more general structures to be defined, and are commonly implemented via either pointers or s-expressions.

The ADS supports a range of type constructors which include i) to v) above and primitive scalar types which include ennumerated types, numeric types and strings. Recursive definitions are supported directly and the conventional pointer constructor is replaced by two new constructors which provide all of the power of pointers without many of the associated problems.

The first of these new constructors is called *STRICT* and adds a *bottom* element to the set of values which constitute its base type. Thus, if *t* is a type containing the values *v1, v2..., vn*, then *STRICT t* is a type which may take the values *v1, v2..., vn* and *bottom*. This is analogous to the way in which pointers may either refer to a value of some type or refer to no value (nil), but without the concept of sharing being involved. Strict types may be used to describe tree or LISP like data structures which have no shared subtrees.

In order to share structures, a type constructor called SHARED is provided which gives sharing in a more controlled fashion than do pointers. The semantics of shared objects, including the concept of reference to them, allow possibly cyclic, hierarchical data objects to be implemented and garbage collection to be organised via the use of reference count techniques[5].

Objects of any type supported by the ADS are stored in a virtual memory which may be distributed over a wide range of memory devices. As these may include non-volatile memory devices which endow data objects stored in them with the attribute of persistence, it is important that the type information associated with a data object is itself persistent. If this were not so it would not be possible to maintain objects in a consistent state by the use of strongly typed interfaces. A positive benefit of maintaining the type information along side the value of a data object is that it may be used to choose automatically a suitable memory representation for the object, according to what kind of memory it is stored in. If the object, or components of it, are moved from one kind of memory to another, the representation can also be modified automatically. Indeed, if the type information is not available it is not possible to move complex objects, specifically shared objects, from one kind of memory to another with any degree of safety.

The ADS's type domain is sufficiently powerful to allow it to describe itself in a straight forward fashion. Therefore types within the ADS are merely special cases of values and may be stored in the same way. In particular, they may be as persistent as the values whose structure they describe. In practice, the type of a complex data object is only stored once and all operations in the domain of values have duals in the domain of types[4]. This allows all objects and their components to be treated as type and value pairs without the overheads of maintaining type information, or even a reference to type information, with each component.

Although the *machine perceivable* memory representations of stored data objects may be deduced automatically from their types and the characteristics of the memory device in which they are stored, it is not possible to deduce representations suitable for *human perception* in the same way. Such representations depend on more complex factors such as who wishes to perceive the data object and why. For this reason the ADS provides a mechanism for defining data object representations suitable for human perception. Currently such representations are all visual representations but in principle they could make use of any human sensory mechanism for which a suitable interface can be made available. e.g if a data object represents a piece of music, the user might want to listen to it or see it in the form of conventional music notation.

The information which describes the *views* which the user may take of an object is again stored in parallel with its value using the same techniques as for type information. The next section describes the form that this information takes.

12.3 VIEWS: CONCRETE REPRESENTATIONS OF ABSTRACT OBJECTS

Abstract types are described in terms of a set of primitives and a set of construction operations. The ADS adopts a similar approach for the description of concrete representations of abstract objects. Each type may have a set of named *view expressions* associated with it. By combining a *view expression* with a value of the type, the ADS is able to form a medium dependent *image* of the value. At any time the user may name the particular *view* which he wishes the ADS to use for the production of subsequent *images*.

The set of *view expression* primitives provided are dependent on the medium in which representations are to be constructed and can be divided into two groups: constant representations which are independent of any stored information and parameterised representations which reflect the state of stored values. The set of construction operations consists of a generic operation for forming compositions of view expressions and media dependent functions of such compositions.

The overall structures of the views associated with a particular type do not have to be related to the structure of the type itself. e.g one view of a binary tree might always yield an image in the form of the string of characters "BINARY TREE", regardless of its actual value. Another view might yield an image composed of a series of nested boxes, one for each node in the tree as below.

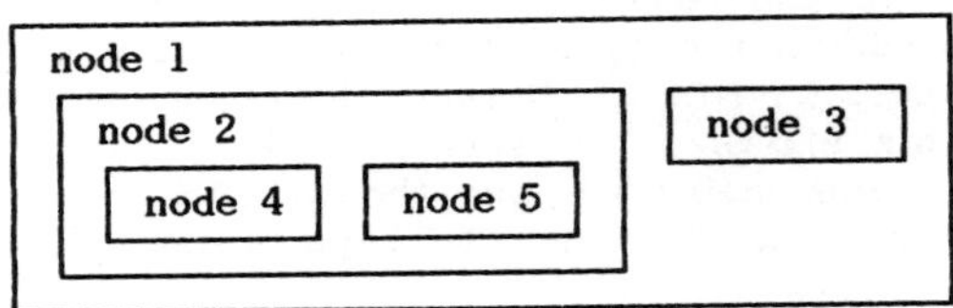

Despite this it is important that the user interface should allow the user to refer directly to stored values via any of the possible representations which the viewing mechanism may produce. The ADS provides this facility and in conjunction with the available type information, uses it to constrain the operations which the user may perform on any component of his stored information. e.g. The user may *point* at any box which represents a binary tree in the diagram above or at the character string "BINARY TREE", and the ADS will be able to determine the type of the object concerned and restrict the operations which may be performed on it to those which are appropriate to binary trees.

The current version of the ADS provides visual representations in a medium which is assumed to be a Euclidian plane or, in terms of abstract data types, a 2-dimensional array of visible components, where the type of a component represents the "character set" of the medium and may for example include other properties such as colour etc. The visual representation of a scalar type value occupies a rectangular region of this plane whose extent and individual component values may be determined by the value of the scalar. Thus for example, a natural number may be represented by decimal digits or by a line whose length is proportional to its value. The visible representation of a composite object is defined in terms of a chosen juxtaposition of a chosen representation of its components, together with selected annotation. Thus the elements of an array structured value could be displayed across the page separated by commas, or tabulated vertically one to a line and preceded by their index. The same facilities apply to the remaining type constructors.

In order to achieve this formatting, it is necessary to provide some default representation for values of scalar types; to allow composition of components above, below, to the left and to the right; and to use constant representations as annotations to bracket composite values and to separate their components.

12.4 EXAMPLE PROTOTYPE USING THE ADS

This section illustrates, by means of a small example, how the ADS has been used. The application chosen is computer aided instruction. A type is defined for the concept of a 'lesson', instances of which can be constructed by a teacher and used by pupils. This illustrates:-

i)   The way in which a computer aided learning package for use by teachers can be constructed as an abstract data type and interactively prototyped to suit the style of teaching and monitoring required.

ii)   The way in which a teacher can prototype lessons by constructing values of the type.

iii) The way in which different views of a structured value can be presented on the one hand to a teacher during construction of a lesson value and on the other to a pupil exploring the value during a lesson.

iv)   The tactile nature of the user interface as it is used to explore and manipulate structures.

The design of the type, which describes the lessons developed by the teachers using the system, assumes that all lessons consist of repeatedly presenting the pupil with information and eliciting a response. The amount of information presented at a time must be appropriate to the pupils comprehension, might be as little as a single sentence or picture, and is unlikely to be more than a page. The response is used as feedback to determine the future progress of the learning process, whether to move on to a new concept, or to repeat a previous concept in the same or a different way.

In ADS terms, the following types capture the ideas above.

```
lesson          = MAP OF frame

frame           = SHARED frame def

frame def       = RECORD
                    title        : line
                    information  : text
                    continuation: lesson
                  END

line            = STRING [60]

text            = LIST OF line
```

A lesson consists of a set of zero or more frames. Many frames may refer to the same frame definition and so frame definitions are shared objects. Each frame definition has a title which may be used to identify it within the lesson to which it belongs. In addition, a frame definition contains some information and a continuation lesson. The title is simply a line of text and the information is represented by a piece of text which is simply a list of lines.

The information is that which is to be imparted to the pupil, who responds by selecting a frame from the continuation lesson. The student may select a continuation lesson in order to answer a question posed in the information, or the selection may represent a free choice of teaching material or merely readiness to continue. The teacher may control the course of the lesson by his choice of continuation lessons, which may present new information to the student or lead him back to review old material, perhaps in a different way.

Therefore, the operations which a data structure of this type must support include the following. The teacher must be able to insert new frames into lessons and delete unwanted frames. He must be able to create and modify frame definitions and bind existing frames into continuation lessons. The student must be able to select frames out of lessons.

Clearly the type is independent of the subject matter being taught, of the precise teaching approach, whether rote learning, practice of techniques etc and of the level of teaching from infant up to higher education; the type is applicable to a wide range of lessons. A value of the type represents a particular lesson in a particular subject, constructed by a particular teacher. Each different teacher can use the same type to construct values for each of his lessons.

Although values of the type contain all the information needed in order for the teacher to express what the pupil is to experience, the individual views which the teacher and pupil have of this information is different. The pupil sees the lesson as a sequence of frames, one at a time, some of which may be repeated in the sequence if incorrect answers have been given. For each frame, the pupil needs to see only the information which is supplied by the teacher and the set of continuations from which to choose. On the other hand the teacher may need to take a more global view of a lesson so that he can reason about the way in which a set of connected frames guide the student through a particular part of a subject.

Thus in constructing a computer assisted teaching tool, in addition to the type, lesson, it is necessary to supply two views for the type and its constituent components. The teacher's global view and the localised view of a frame definition which will be used by both student and teacher.

Prototyping the above system using the ADS starts with the creation of the type to represent lessons. Although such a type has already been described in this section, the design of a type is usually performed interactively when using the ADS. This gives the designer access to facilities which allow him to focus on as much or as little of the detail of the type structure as he wishes and to take advantage of facilities which support stepwise refinement of a type and allow automatic cross referencing between its components.

As has already been stated the ADS type domain is described in terms of itself. Thus, when a designer is creating or modifying a type, he is working with a value, the type of which is defined in the same type notation. Therefore he only needs to learn one notation in which to express his ideas *and* to be able to interact with the system. This should be contrasted with conventional programming environments in which programming languages are described in notations such as BNF and users need to learn a meta notation and a (different) programming language notation before they can interact with a computer system effectively.

The rest of this section attempts to convey some 'feel' for what it is like to use a system organised in this way. However, the ADS interface is an extremely dynamic one and to date the authors do not feel at all confident in their ability to do it justice in a description which must be written down statically on two dimensional pieces of paper.

The following diagram shows a view of the lesson type as the designer might see it at an intermediate stage in its construction via a display connected to the ADS.

```
Change VARIANT(<type defn>):   <- ->   scalar shared string map list


   TYPE lesson            = MAP OF frame

         frame            = SHARED frame def

         frame def        = RECORD
                                title           : line
                                information     : text
                                <identifier>    : <type>
                            END

         line             = STRING [60]

         text             = -32768..-32768
Enumeration name ?
```

At this late stage in the design of the lesson type the designer is
making a decision about what kind of object the type text represents.
He has told the ADS that he wishes to change the highlighted type
definition and the ADS is asking what kind of type he would like to
change it into. If he types 'list' to the prompt at the bottom of the
screen or points at the highlighted button marked 'list' at the top of
the screen the display will change to that shown below.

```
<type defn>    <- ->  V(iew f(orward b(ack T(ype C(hange S(tore R(ecall


   TYPE lesson            = MAP OF frame

         frame            = SHARED frame def

         frame def        = RECORD
                                title           : line
                                information     : text
                                <identifier>    : <type>
                            END

         line             = STRING [60]

         text             = LIST OF <type>
```

It should be stressed at this point that the interface to this type
description is generated by the ADS from the information which it has
about the type of types. The lesson type is merely a value of this
type. What the designer sees on his display is one of several available
views of a value of the type of types. It is generated in exactly the

same fashion as views of user defined types. This particular view looks just like the conventional representation of a type as a piece of text. The disadvantage of conventional textual representations is that they only support directly operations which are relevant to sequences of characters. The ADS representation supports operations that are relevant to the types which the user wishes to manipulate.

If in the previous example the designer had typed 'apple' in response to the 'Enumeration name' prompt, the ADS would have refused to change the selected type definition as the type domain does not include an 'apple' type constructor. The user can point at any character on the display and the ADS will know exactly what component of the displayed value it is a part of, what type of component that is and consequently what operations the user is allowed to perform on it.

As a further example of the difference between this representation and a piece of text, consider the view of the type definition for the type 'frame'. On the right hand side of the equals symbol it reads 'SHARED frame def'. The string 'frame def' might appear to be just the name of the shared type being defined. However, in this representation it *is* the type called 'frame def'. If the designer were to change the name 'frame def', wherever it occurs on the display, all the other occurrences would change as well. This is because there is only one copy of the type name in the system, it is just displayed in more than one place on the display by virtue of the ADS generating, in this case, two different views of the type definition of which it is a part, i.e. the representation of the type seen on the two dimensional display has the same multi-dimensional properties as the type it represents.

When the user wants to specify the type of a component of some other type, he simply points at the component to be defined, pushes the button at the top of the screen marked 'change' and then either points at the type he wishes the component to become, if such a type is on the display, or types the name of the required type. If the type already exists the ADS will construct a new reference to it, otherwise it will construct a new type, add it to the display and then construct the reference to it. If the new object is ever destroyed, all references to it will automatically revert to their undefined state.

Once the type which models the application has been defined, further information can be associated with each component type in order to define one or more views of values of the type. If no views are defined the ADS will supply default views based on the type constructors employed.

A view expression may be thought of as defining a function which may be applied to a value of the type with which it is associated to form an image of the value. Each type definition may have any number of views associated with it and a view expression may identify specific views which are to be taken of its components. Such sub-views are identified by name and are implicitly associated with the type of the component concerned. The following example demonstrates the functional nature of the view descriptions.

```
lesson(student)      = down(*frame(name))

frame(name)          = frame def(name)

frame def(student) = down(enclose(down(title(student)
                                        +
                                        information(student)
                                        )
                                   )
                          +
                          enclose(down('Continuation...'
                                       +
                                       continuation(student)
                                       )
                                  )
                          )

frame def(name)      = title(name)

line(student)        = centre justify(60)
line(name)           = left justify(0)

text(student)        = down(*line(student))
```

The first line of this view description indicates that there is a
'student' view of a lesson which is composed of all the frames in the
lesson map laid out down the display. Each frame is to be seen in its
'name' view. The 'name' view of a frame is simply the 'name' view of
the frame definition to which it refers. The 'name' view of a frame
definition only contains its title and nothing else. However, there is
also a 'student' view of a frame definition through which the student
will generally interact with it. This is a composition of two components
arranged down the display. The first is an enclosed view of the title
arranged above a view of the information contained in the frame
definition. Enclosed components have boxes drawn around them. The
second component is an enclosed view of the string 'Continuation...'
arranged above a view of the continuation lesson.

In practise a little more information must be added to such a view
description to define views for special cases such as empty lessons and
undefined shared values. This is achieved by associating views with
special default names with the types which require special views for
such special cases. e.g. If the ADS is asked to display an empty map it
will disregard the prevailing view it has been told to use and attempt to
use a view called 'empty'. If no view with this name is associated with
the map type concerned it will use a system dependent default view.

Completed view expressions may be combined by the ADS with values
of their associated type to produce images on a display. The following
picture shows an image that a student might see after selecting a
particular frame definition out of a lesson. It is assumed that the
lesson value has already been constructed by a teacher and that the
ADS has been told to used the 'student' view.

```
<frame def>    <--> V(iew f(orward b(ack T(ype C(hange S(tore R(ecall

              Identifier Syntax

    Which of the following BNF definitions most accurately
              describes a Pascal identifier?

    1) <identifier> ::= <letter> <identifier> | <digit>

    2) <identifier> ::= <letter> <rest>
       <rest> ::= <letter> <rest> | <digit> <rest> | <empty>

    Continuation...

    Answer 1
    Answer 2
```

The student has only to point at one of the lines in the box headed
'Continuation...' and the display will change to show the frame selected
from the continuation lesson. The 'student' view information for a frame
definition specified that the 'student' view was to be used for the
continuation lesson and this in turn specified the 'name' view of a
frame. This accounts for only the titles of the frames in the
continuation lesson being visible on the above display. However, when
the ADS is told to display the frame selected from the continuation
lesson it will automatically dereference the frame to get to the a frame
definition and then display that in the prevailing 'student' view. Thus
the selected frame will appear on the display in the same amount of
detail as in the example above.

Obviously the teacher designing a lesson will want to see the
'student' view of it, but will also need to see a higher level
representation in which frames can be seen in context but not in detail.
An alternative view called a 'teacher' view might be constructed to allow
this.

```
lesson(teacher)     = enclose(right(*frame(teacher)))


frame(teacher)      = frame def(teacher)


frame def(teacher) = down(title(name)
                          +
                          continuation(teacher)
                         )
```

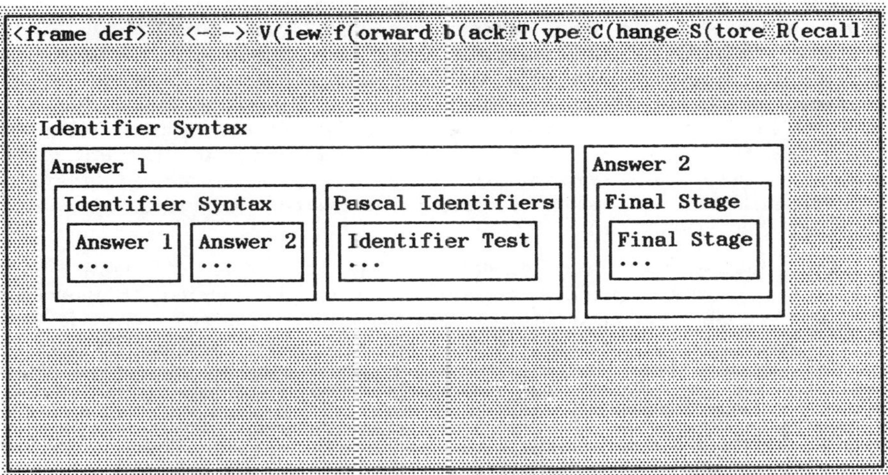

This 'teacher' view of the same frame definition clearly shows that there are cyclic paths within the example lesson. The above display shows the result of associating a low visibility level with the view of a lesson. This has resulted in a special, but still user defined, view of a lesson being used when it is displayed more than five levels below the root of the display. However, the ADS can still be told to display more detail just by pointing at the '...' which represents this special view of a lesson. The next five levels would then be displayed without any missing detail. If the visibility of a lesson were not limited in this way, the ADS would simply clip the image at the display boundary. This avoids the need for an infinite display for such values!

This example has only illustrated some of the facilities of the view mechanism of the ADS. Others allow for the protection of components of a value so that, for example, students would not be allowed to select and change the text or structure of a lesson and the teacher would not be allowed to change its type.

From the prototyping point of view, a teacher developing learning material does so by editing the value of his variable of type lesson. He can change independently the presentation of the information, the questions and their choice of answers and the order in which the material is presented, simply by selecting the component he wishes to modify and supplying its alternative value. Similarly, an educational supplier marketing computer assisted teaching packages, would adjust the type to meet the educational needs. For example, a more sophisticated package which monitors and logs a pupils performance can be constructed by expanding the types provided to include one for a log which consists of a list of cross references to responses, onto which each pupil selected response is appended. Alternative views of the log are capable of providing a teacher with different levels of detail about a pupil.

12.5 CONCLUSIONS

The ADS provides a well structured, strongly typed persistent data store which extends and unifies concepts from hierarchical filing systems, data bases and strongly typed languages. In addition the viewing mechanism allows user defined concrete representations of its contents for human perusal. The power of the viewing mechanism can be used in the same way as sub schema in data bases, to allow different users selective access to the data or simply to view it in different ways. e.g. A program value can be viewed in a high-level source language representation or as a piece of assembly language, eliminating completely the need for compilation.

The design technique employed with the ADS, of using data structures to model problem domains is of course familiar to users of design methods such as JSD[6] and Data Structured Design[7]. What is new and of particular significance is the ease with which the data types can be set up and the clarity with which they can be presented to a customer in the prototyping process. The ADS has thus proved appropriate for top down refinement of prototype designs, which can readily be modified as a result of user reaction to the appearance and behaviour of the system prototyped. Implementations are realisable by further refinement of the prototypes. Thus, the techniques of top down, stepwise refinement can be applied equally to the processes of prototyping and design of software systems and, as indicated below, to their specification and implementation.

The system is still evolving and a number of further developments are already planned. These fall into two classes, extension of the user facilities and modification of the implementation. The former include extension of the type concept to that of classes, by inclusion of defined operations within a type and subtypes defined by constraints which can be used for specification and consistency checking. These extensions will widen the area of applicability of the ADS to include the more formal aspects of specification. Alternative user interfaces are also a planned area of investigation. On the implementation front, the provision of persistent mutual exclusion on persistent data will permit multiuser and multiaccess facilities, which are a natural consequence of a fully distributed implementation and which further extend the range of applicability of the ADS.

The novel concepts involved in the ADS have emerged in parallel with its development in a highly experimental manner, which supports the thesis that prototyping is a valuable system design tool. The very lack of such a tool during its development has emphasised the importance of speed and flexibility to prototyping. With the planned extensions, the ADS will provide a powerful and flexible integrated project support environment suitable for both the distributed development of software and the development of distributed software. In this environment *programs* will merely be special cases of data structures and the system itself will be both operating system and machine architecture independent.

ACKNOWLEDGEMENTS

The work described here was partially supported by SERC research grant nos. GR/B35062 and GR/C34779 and Cooperative award no. GR/B78861, in cooperation with BL Systems Ltd. The authors would also like to thank Graham Binks, Chris Tan, Tony Tye and Martin Williams for their suggestions and contributions to the work described.

REFERENCES

1. Atkinson M.P., Chisholm K.J., and Cockshott W.P.
   "PS-Algol: an Algol with a Persistent Heap"
   ACM Sigplan Notices 17(7), 24-31 (1982)

2. Atkinson M.P., Chisholm K.J., and Cockshott W.P.
   "An Approach to Persistent Programming"
   The Computer Journal, 26, (4), 360-365 (1983)

3. Hoare C.A.R.
   "Data Structures" in "Structured Programming"
   Dahl O.J., Dijkstra E.W. and Hoare C.A.R.
   A.P.I.C. Studies in Data Processing No. 8.

4. Hughes J.W. and Powell M.S.
   "A Strongly Typed Distributed Virtual Memory"
   in "Distributed Computing Systems Programme"
   Ed. D.Duce, Peter Peregrinus Ltd., 1984

5. Hughes J.W. and Powell M.S.
   "Program Development Using an Abstract Data Store"
   Submitted for publication, September 1984

6. Jackson M. A.
   "Information Systems: modelling sequencing and transformations"
   IEEE Proc. International Conference on Software Engineering, 1978.

7. Knight C.
   "Data Structured Design Methods"
   Advances in Data Structured Design,
   Infotech State of the Art Seminar, 1980.

8. Liskov B., et. al.
   "Abstraction Mechanisms in CLU"
   CACM, Vol 20, No. 8, 1977.

9. Wirth N.
   "The Programming Language Pascal"
   Acta Informatica, Vol 1, pp 35-63, 1971.

Chapter 13

IPSEs in commercial data processing

P.W. Sellars

13.1 INTRODUCTION

This paper summarises work done on the definition of
an integrated project support environment for use in
commercial data processing. It indicates the requirements
and design objectives for such a system, outlines the
target environments with which the support system must
co-exist and describes the principal features required.

The requirements of a support environment arise from
the following aspects of system development activity:

- satisfaction of end-user requirements,

- maintainability of systems,

- control of system development activity,

- cost-effectiveness of staff activities.

The degree to which end-user requirements are
satisfied by the application systems developed depends on
many factors, most of which are not directly influenced by
provision of a support system. The factors which are
significant are quality of specification, opportunity for
user-involvement, meeting of deadlines and control of
costs. The associated requirements are the following:

- enable accurate and consistent specifications to be
  produced to a good standard of legibility during each
  stage of review and agreement,

- aid the planning and progressing of projects,

- aid the estimating and control of costs of both
  development and system operation.

The maintainability of an application system depends
on the quality of the design but also on the reliability
and accessibility of system documentation. The associated
requirements are the following:

- provide facilities to assist in the checking of
  completeness of documentation,

- aid the checking of consistency of documentation,

- aid the assessment of the impact of proposed changes
  to systems,

- provide ready access to system documentation, both to
  specific items and to sets of related items,

- encourage the maintenance of documentation by
  production of action lists of impact of changes.

The control of the system development activity is of
the following aspects: authorisation and sign-off,
resources consumed, quality of product and adherence to
standards. The associated requirements are the following:

- provide facilities to aid in the control of project
  authorisation and sign-off and in the monitoring of
  work done and resource consumption,

- aid the quality assurance function by allowing easy
  inspection of documents produced, highlighting of
  omissions and checking of consistency,

- provide ready access to standards documents,

- encourage the observance of standards by incorporating
  their features within the facilities.

The extent to which staff are used in a cost-effective
manner depends on the possibilities for: removing tasks
which do not require their particular skills, aiding the
organisation of work done and the provision of tools to
increase productivity. The associated requirements are the
following:

- provide automated facilities to carry out repetitive
  or routine tasks,

- provide a framework for the organisation and retrieval
  of all system documentation,

- provide tools to assist each major type of system
  development task.

13.2 DESIGN OBJECTIVES

The design objectives for a support environment
are as follows:

- the system should be capable of being used in
  co-operation with a variety of existing automated
  aids,

- the system should be capable of being used to support a variety of working practices and development methodologies,

- the system should be capable of being used with a variety of target environments (ie the hardware and software environment for which an application system is being developed),

- the design should take account of the fact that the scope of the environment will be reviewed frequently during its life,

- each facility should be capable of use either independently of or in co-operation with other facilities,

- the system should be capable of use by all types of system development staff and, for bulk input, by typists,

- the system should encourage rather than impose the observance of standards,

- all automatic functions should be capable of manual override,

- the system should be 'forgiving' in the sense of providing easy recovery from unintended actions,

- the style of use and the presentation of each facility should be uniform throughout the system,

- the use of the facilities should involve sensible procedures and not impose overheads on project work,

- explanations of how each facility is used should be directly available in the form of 'help' screens accessible while using the facility.

13.3 TARGET ENVIRONMENTS

The system must be capable of operating within a wide variety of environments determined by hardware, software and working practices.

13.3.1 Target Machines

The support system must communicate with the target machine for the following purposes:

- submission for compilation of source code developed on the support system,

- retrieval of data descriptions held on the target
  system in a data dictionary or a source code library,

- submission of data descriptions to the target system,

- use of services provided only on the target machine.

The operating system and job control language on the
target system are of significance in determining the
facilities provided for assembling jobstreams on the
support system and for identifying ownership of output from
the target system.

Use of the support system can be independent of the
particular programming language selected for an
application. This is because the support system is not
itself concerned with producing object code. Hence, the
minimum level of use of the support system treats source
code simply as a file of text.

13.3.2 Database Management Systems

The type of Database Management System to be used in
the target environment influences the use of the support
system in the following ways:

- the method of producing and refining the database
  design,

- the choice of primary repository for the database
  definition,

- the provision of macros for database calls.

Where a data dictionary exists in the target
environment, it is used in conjunction with the support
system. Data dictionaries are used in a number of
different ways depending on:

- facilities available within the software,

- Database Administration approval procedures,

- intended coverage of the dictionary in terms of types
  of data to be recorded.

The facilities vary in terms of the degree of support
provided for logical and physical descriptions, the ability
to support all categories of documentation and the extent
to which different versions and statuses of data
description can be maintained.

Database Administration approval procedures vary
depending on the prominence of the dictionary within the
installation and whether the dictionary is used to support
projects individually or corporately. The approval
procedures also depend on the ability of the dictionary to
support different versions and statuses of data.

The intended coverage of the dictionary varies depending on the stage of development of the installation and on policy decisions on categories of application to be supported.

The above factors influence the extent to which the data dictionary or the support system becomes the primary repository of data and, for each category of data, which system is regarded as the master and which is the slave.

13.3.3 Source Code Development

Where, as is commonly the case, interactive programming facilities are available on the target machine, source code can be developed in a number of ways:

- initial development of source code on the support system with subsequent transfer to a source code library in the target environment,

- use of the support system as the primary repository of source code with freeing of the mainframe facilities for testing activities only,

- use of the support system only for the phases up to and including logical program design with source code development making use of the existing facilities.

13.3.4 System Development Methodologies

The support system embodies the principles of formal system development methodology but within those principles, there is some flexibility as to how the support system is used. The significant principles are the following:

- projects should be planned, conducted and controlled in terms of defined phases and tasks,

- tasks should be associated with defined items of documentation,

- documentation should be defined in terms of discrete items each of which is a component of some level of specification,

- relationships between items of documentation should be well-defined,

- where diagrams are used, these should employ standard symbols and have well-defined relationships to supporting documentation.

13.4 FEATURES OF THE SUPPORT ENVIRONMENT

The support environment provides facilities to assist in carrying out the following functions:

- Document Production,

- Design Verification,

- Project Library Maintenance,

- Document Viewing,

- Standards Viewing,

- Performance Modelling,

- Project Management,

- System Administration,

- Use of Mainframe Facilities,

- Prototyping.

Facilities are invoked via a hierarchy of menus. Different categories of user are provided with a hierarchy appropriate to the facilities they require, or are permitted, to use. From each menu, a 'help' option may be selected to provide explanation of the facilities available at that level. In addition, help options are associated with the detailed operation of each facility.

Most of the facilities are concerned with processing of documents. The system automatically organises documents by type within project and controls access to the documents. During development of a document, the author is given exclusive permission to make changes. Once approved, the document is located in the Project Library. Access permissions are changed so that amendments may then only be made by nominated individuals.

13.5 DOCUMENT PRODUCTION

All documents are created and amended using a full-screen editor. In the case of reports, the facilities are provided by means of word-processing software. For other documents, simplified editors may be used having reduced functionality but also some degree of knowledge of the rules governing completion of the document.

The precise type of editor to be used for any particular purpose is capable of being varied depending on the skills of the user. For example, bulk input of documents is normally achieved by typists using word-processing software. Systems Analysts with less developed keyboard skills may prefer a simpler editor.

From the user's point of view, individual documents
are normally retrieved by requesting a picking list of
descriptive names in some category and selecting the name
required. The selection is performed either by entering a
sequence number or by moving the cursor to the name
required. All documents are further described by items
indicating authorship, application, status and version.
Most types of document have some defined structure and
make use of standard headings or other portions of text.
These serve as an aide-memoire to the user and also help to
impose a standard style on documents produced. For each
such document type, one or more skeleton documents are
stored and may be selected for use as the basis for
creation of new documents. The skeleton documents are
defined when producing the system development standards for
the client and are stored in the Standards Library.
Portions of some documents can be re-used between one
phase and a subsequent phase or between one project and
another. One case is where common documentation is stored
on a mainframe data dictionary. A copy of the items
required is retrieved for reference or further development
with the current project. Other examples of the re-use of
documents are the following:

- the progression of management reports produced at the
  end of each project phase where certain text may
  remain constant between one report and the next,

- procedure specifications produced during the Analysis,
  Design and Programming phases where there is a
  progression of level of detail but overall logical
  structure may remain constant.

Each type of document can be viewed on the screen or
printed. For viewing, the following options are provided:

- scrolling vertically and horizontally,
- split-screen viewing of portions of several documents,
- cursor selection of other documents referred to in the
  text or represented by a diagram symbol.

For each type of document, facilities are provided for
copying, renaming and deleting documents. Documents may be
copied or moved between the following locations:

- within the user's work area,
- between one user's work area and another,
- from the Project Library (security is imposed on
  copying to the Project Library),
- to and from mainframe files (using RJE),

13.6 DOCUMENT CATEGORIES

Documents are categorised according to the type of processing to which they may be subjected.

13.6.1 Forms

A Form is any document for which standard headings can be prescribed and where each heading corresponds to a concise entry.

Two methods are provided for the input of forms: provision of a skeleton corresponding to the paper document or provision of a dialogue requesting entry of information under each heading.

The first method is normally used for bulk input of documents or where a significant proportion of items are optional. The second method is used for interactive entry of information and is appropriate for forms where most items are mandatory. In each case, the description of the form is stored in the Standards Library and is tailored to the particular requirements.

Certain of the headings on each form are significant in the sense that they correspond to cross-references with other documents. Hence there are certain restrictions on the extent to which they can be modified.

13.6.2 Reports and Specifications

A Report or Specification is a document assembled from other documents. The component documents may be produced by one or more people. There is generally a requirement to edit the complete report before its final production. For standard reports, the required sections are listed in the Standards Library together with details of page headings and footings and any other standard portions of text. When working on a report, the standard structure is copied into the user's work area. The structure may then be varied to that required for the particular report. Each person working on the report is provided with the overall structure and may in principle be working on any of the sections.

The report is eventually collated by copying the component documents to the Project Library and then merging and editing then as required. At this stage, facilities are provided to define subsets of reports for particular audiences.

13.6.3 Procedure Specifications

A Procedure Specification is a document specifying the rules to be followed for carrying out a procedure. It is written using the logical constructs, Sequence, Selection and Iteration and refers to defined data items. Such specifications are produced during the Systems Analysis, System Design and Programming phases.

Procedure Specifications are produced using an editor

as for any other kind of text. Once produced they are
processed to achieve the following:

- checking syntax rules for logical constructs,
- formatting of text to highlight logical constructs,
- identification of data items to allow cross-checking
  with data definitions.

Provided that Procedure Specifications make use of the
prescribed logical constructs, they can be processed as
described. The particular keywords used to identify the
constructs are defined in the Standards Library so that
various conventions can be accommodated.

13.6.4 Source Code

Source Code is any compilable or machine-interpreted
language. The facilities for handling COBOL or PL/1 are
essentially the same as those for 4th Generation languages
or Job Control languages.
The facilities specific to the handling of source code
are as follows:

- A selection of skeleton programs may be set up in the
  Standards Library to cater for commonly required types
  of program in the appropriate languages.

- Macros may be set up in the Standards Library to
  provide commonly required portions of code. The
  macros are coded making use of symbolic names for
  which the desired substitutions are automatically made
  when including a macro within a program.

- Design Language logical constructs which are defined
  in the Standards Library may be associated with their
  equivalents in the various target source languages.
  Facilities are provided to use these to generate a
  skeleton program from the corresponding Program Design
  Language.

- 'Shorthand' forms of source code keywords may be
  defined in the Standards Library. These allow source
  code to be written in an abbreviated form and then
  automatically expanded to conform with the language
  rules. The same mechanism can also be used for
  shorthand forms of any character strings to be
  included in a program.

- Syntax checking of source code is provided by using
  local compilers. This facility is restricted to those
  languages for which compilers are available on the
  machine (COBOL, FORTRAN, PL/I) and is limited by the
  versions of the compilers. (In general, compilations
  to produce object code must be performed on the target
  machine).

13.6.5 Test Conditions and Data

The identification of conditions to be tested when planning Program, System or Acceptance Testing is in part conducted by inspecting the corresponding level of specification of the system. Facilities are provided to generate an initial framework for the identification of test conditions. The framework is generated by scanning the procedure specifications and extracting the names of conditions to be tested. The framework provides the initial version of the test conditions document which is then developed using an editor as for any other kind of text.

Test Data may be produced manually or generated according to specified rules. To some extent, automatic generation can take place from conditions previously identified.

13.6.6 Diagrams

Diagrams are produced on conventional screens using a character set containing the primitives of the symbols required. Hence the same low-cost screens are used for both text and diagram processing.

Diagrams are of two types according to their method of production:

- those which are produced by the user and input to the system,

- those which are produced by the system by processing or analysing other items of documentation.

In the first category are diagrams such as Data Flow Diagrams or Data Structure Diagrams. These are diagrams which are generally produced prior to the production of their supporting documentation. Also included in this category are those diagrams which, while they could be produced by analysis of other documentation, raise problems of symbol positioning. In the second category are diagrams such as Bar Charts or Critical Path Networks. Diagrams are constructed from symbols whose shape and meanings are stored in the Standards Library.

Production of the diagram involves the following steps:

- position cursor at required location,

- select symbol from menu,

- optionally adjust size and position of symbol,

- enter text, naming and numbering the symbol.

The screen acts as a window on a larger work space. The normal method of working is to construct the diagram so that it falls within one or more A4 sheets (to facilitate production of hard copy). Facilities are provided for reducing the diagram so that it can be viewed as a whole (though without the text descriptions).

As each symbol is positioned on the diagram, its type, location, name and number are recorded in a diagram description. The diagram description is used for reconstruction of the diagram and also to provide the basis for generation of skeletons of the documentation necessary to support the diagram. The skeletons are stored in the same way as the documents themselves but are given a status to indicate that they are incomplete. The skeletons can be completed immediately after diagram production or on a subsequent occasion. The diagram description ties the diagram to the supporting documents including those which exist prior to production of the diagram.

Changes to a diagram potentially result in changes required to the supporting documentation. These potential changes are presented as an action list to the user.

Diagrams can be read in conjunction with their supporting documentation by positioning the cursor on a symbol and using a function key to request display of the corresponding document. In the case of hierarchies of diagrams the same mechanism is used to view the next level of diagram.

13.7 DOCUMENTATION MANAGEMENT

13.7.1 Project Libraries

A Project Library is maintained for each project conducted on the system. It is the repository for all documentation which has been completed and approved.

The following facilities are provided for maintenance of a Project Library:

- Access control preventing unauthorised updating,

- Selective document copying from User Work Areas,

- Version control,

- Consistency checking,

- Index production,

- Releasing of documents for change.

Documents within the Project Library may be viewed by any member of the project group. Changes to the Project Library can only be made with appropriate authorisation. No facilities are provided for direct change to the Project Library except by copying to and from User Work Areas.

Action lists are generated of documents which have been completed and approved. The list is displayed and used to prompt for decision on whether to move to the

Project Library.
     Documents within the Project Library are located
within directories corresponding to the type of document.
Hence, for reports, the process effectively collates the
sections, which may have been produced by a number of
individuals, into a single report directory.
     Facilities are provided for the maintenance of
multiple versions of documents stored within the system.
Different versions of documents arise in two ways:

- from the progression of documents during a project or
  the life of the system resulting from changes or
  correction of errors,

- from the need to produce parallel versions of a system
  to run in different environments.

     Version control is exercised automatically on
documents within the Project Library. Optionally it is
available for use within an individual user's project work
area. It enables recovery to a nominated version and the
identification of changes since a nominated version.

13.7.2 Document Relationships

     Each item of documentation in the Project Library may
have one or more relationships to other items of
documentation. Two broad categories of relationship are
distinguished, one being the inverse of the other:

- 'content' relationships where the content of a
  document refers to other documents,

- 'usage' relationships where a document is referred to
  by other documents.

     For each document type, a list is maintained of the
relationships in which the document may participate. This
list is created when defining document types to the system.
     For each document in the Project Library, a Content
List is generated. The Content List contains an entry for
each relationship which the document has. The Content
Lists are inverted to generate or augment Usage Lists for
the documents referred to.

13.7.3 Design Verification

     Design verification is concerned with ensuring that
the set of project documentation is compete and consistent.
The facilities provided cannot prove the completeness or
consistency of a set of documentation but do aid in the
routine aspects of checking.

Completeness is checked by ensuring that:

- all expected items of documentation have been
  produced,

- all compulsory sections of a document are present,

- document dependencies between phases are satisfied
  (e.g. processes for each function, physical data
  items for each logical data item).

Consistency is checked by ensuring that system
components referred to in one document are themselves
defined. The facilities are used to aid the review by a
project team of project documentation and also to assist in
carrying out Quality Assurance.
The facilities are concerned only with reporting
apparent discrepancies; the system does not override the
judgement required for approving a set of documentation.

13.8 PERFORMANCE MODELLING

Performance modelling provides for the estimating of
system performance at various key points within a project.
The initial building of a model can be achieved either
by explicitly entering and storing the information required
or by extracting the relevant information from documents
already stored. The facilities make use of a library of
performance parameters for selected target environments.
A performance model has the following components:

- Performance parameter library,

- List of processing components,

- List of stored data, inputs and outputs,

- Definitions of the hardware and software resource
  types used by a processing component,

- Definitions of data volumes,

- Definitions of the number of units of
  hardware/software resource required for a single
  execution of each processing component,

- Definitions of the frequency of execution of each
  processing component.

Each of the above components can be amended in order
to vary the assumptions on which the model is based.
Processing frequencies can be stored by time period to
allow the model to be run separately for each.

13.9 PROJECT MANAGEMENT

13.9.1 Project Planning

The identification of project tasks is aided by the provision of lists of standard tasks per project phase. Standard task lists are maintained in the Standards Library together with estimating criteria for dedicated and elapsed time. Facilities are provided to use these as a start point for project planning. The task list required is copied to an initial version of the project plan and there developed by inserting additional tasks, amending task names and deleting tasks not required.

Estimating is aided by providing ready access to the source information required. The estimating criteria attached to each task are inspected to arrive at the estimates for each task and the minimum and maximum estimate is recorded. Project Size statistics (numbers of activities, processes, inputs etc) can be generated at the end of each phase and inspected during the estimating process.

13.9.2 Project Scheduling

Project Scheduling is aided by the provision of the following facilities:

- recording of dependencies between tasks,

- production of bar charts,

- identification of Critical Paths,

- derivation of effects of changing start and finish dates.

The facilities for Resource Assignment can be used in association with those for Project Scheduling so that the impact of non-availability of resources can be tested. Following Resource Assignments, reports can be produced of resource commitment by, for example, resource type and time period.

Project Progressing is aided by the provision of information on actuals versus estimates and by the ability to test effects of slippage on project schedules. Information can be printed by various categories or displayed at the screen in response to entry of selection criteria.

13.9.3 Project Costing

Costing of project plans is achieved by processing of Resource Assignments. The unit cost of each resource is recorded against the particular assignment to a project task. Costs can be totalled by resource type, project phase and by time period. During project planning, the

cost rates are applied to time estimates and used to assist
in the preparation of project budgets. During the conduct
of the project, the cost rates are applied to the time
spent per assignment and used to generate records of actual
spend against budget. Staff time is directly input by
individuals.
   Project Progressing is aided by the provision of
information on actuals versus estimates and by the ability
to test effects of slippage on project schedules.
Information can be printed by various categories or
displayed at the screen in response to entry of selection
criteria.

13.9.4 Estimating Criteria

   The estimating criteria stored within the system are
capable of modification in the light of experience. A
major input to this is the actuals figures gathered by the
support system. Facilities are provided to express actuals
in a form suitable for comparison with estimating criteria.
The exercise may be carried out on individual projects or
across families of projects. The facilities provided are
as follows:

   - derivation of project sizing information by counting
     numbers of functions, inputs, outputs, units of
     documentation etc.,

   - derivation of ratios of project sizing data between
     one phase and another,

   - derivation of ratios of resource consumption per unit
     of documentation or other indicators of project size,

   - analysis of proportions of time spent on various types
     of task.

13.10 CURRENT DEVELOPMENTS

   BIS Applied Systems Ltd has embarked upon the
development of an integrated project support environment
(BIS/IPSE) based upon the conclusions reached in this
paper. Release 1 has now been developed and is in use at
a number of installations. Further releases are under
development.

Chapter 14

Mentor—Design and implementation of the kernel of a program manipulation system

B. Lang

14.1. INTRODUCTION

Early attempts at building systems for proving, transforming or deriving programs, or even other symbolic manipulation systems, have often bogged down into what has been sometimes called the syntactic quagmire. The development of the semantic parts of these systems was limited or complicated by the inadequacy of the tools available to manipulate whatever representation of programs was being used.

It was then recognized that, in order to build an advanced system dealing with the semantics of programs (and more generally with the semantics of formal systems), one has first to attend the more mundane (and sometimes tedious) task of providing a well engineered collection of syntactic facilities needed to represent and manipulate programs.

The original purpose of the Mentor system was precisely to be such a syntactic facility, with the added purpose that the system should be usable for life-size program manipulations rather than for experiments on small examples. Although the ultimate aim was still to build a semantics based program manipulation tool, this syntactic component turned out to be at the same time sufficiently challenging to become a research project in its own right, and sufficiently powerful to be used as a program development system even without the adjunction of semantic components.

Mentor is not an integrated project support environment (IPSE). Its range of applications goes beyond mere syntax directed editing to a variety of programming support activities (program transformations, transport, preprocessors, documentation, etc.), but it is essentially limited to manipulations of syntactic representations of programs. The systematic integration in Mentor-like systems of tools based on semantics is still the object of ongoing research [1,4,29] and does not seem yet mature enough for an industrial production system. Specific tools (e.g. compilers) could be to some extent integrated on an ad hoc per case basis; we have not attempted this mainly for reasons of manpower and portability. Finally Mentor does not support team oriented activities such as maintenance of a program data base, communications between users or project management.

However, as was initially intended, the concepts and techniques developed for Mentor can provide a sound and uniform basis for all syntactic aspects of an IPSE.

The Lisp implementation reported in the second part of this paper has been partly supported by contracts from the Concerto Project of the Centre National d'Etudes des Télécommunications, and by the Commission of the European Communities within the Esprit Project 348 (Generation of Interactive Programming Environments).

The object of this paper is to present these concepts as originally developed for Mentor in its first implementation in Pascal, to discuss implementation issues raised by the use of this language, and finally to present a new implementation in Lisp including only the kernel functionnalities of Mentor, which is intended to be the syntactic component of an IPSE developed by the Concerto Project of the Centre National d'Etudes des Télécommunications [2].

14.2. THE MENTOR SYSTEM

Mentor is an extensible structured document manipulation system [5,6], intended to serve as a basis for the development of software environments. The main motivation for this approach is the fact that most of the activities of a software project can be viewed as the creation, modification, analysis and maintenance of a variety of related documents: requirements, models of the solution, specifications, code, tests, documentation, user manual, etc... Supporting these activities entails basic choices concerning the following issues:

- the representation of documents,

- the document manipulations primitives,

- the organization of the collection of project related documents,

- the (partial) mechanization of document manipulation activities,

- the design and organisation of tools for the analysis, manipulation or transformation of documents.

14.2.1. Brief history of Mentor.

The implementation of Mentor was started in 1974. The language Pascal was chosen both as the implementation language, and as the object language, i.e. the language of the programs to be manipulated by the system. This choice had the advantage of allowing us to experiment our system on our own programs, and in 1977 we started using Mentor for its own development. In 1981, the first version of the Metal language was designed and Mentor became language independant. Later improvements include an operational full screen interface and extension of the annotation facility.

The system was first developed on a CII 10070 computer, then on a CII IRIS-80, and finally on a CII-HB DPS-8 under Multics. It was tranported at various stages of development on several other machines, and is currently maintained and distributed on DPS-8 under Multics, on VAX-780 under Unix Berkeley 4.2, and on the SM-90 (a Motorola 68000 workstation) under SMX (a version of Unix V7). It is being transported on VAX VMS, on HP 9000, on Data General MV10000 and on the Apollo workstation. It is now running on about 50 sites, of which about 20 are outside France, and it is used in several pilot or Esprit projects.

The current implementation represents approximately 40 000 lines of Pascal code (compiled into about 350 K-bytes of object code), plus several thousands lines written in Mentor's own languages (Mentol and Metal). Mentor itself is used for the source code transformations required by the transports, and the simultaneous maintenance of the distributed implementations.

Unix is a well known trademark of Bell Laboratories.

14.2.2. Representation of documents.

Along with other structure oriented systems [13] Mentor uses a tree structured representation (traditionally called **abstract representation**) of documents instead of the more traditional string representation.

The abstract representation has several advantages:

- It emphasizes the internal structure of documents (especially, but not only, for programs) and supports more effectively their manipulation in terms of this structure. Actually, many traditional tools perform a preliminary syntax analysis to translate documents into a more processable form. This ubiquitous syntax analysis is factored out by using directly the abstract representation.

- It defines a unique standard representation of documents, independant of any concrete physical support and layout, and thus provides a uniform interface between the processors in the environment. One may even have several concrete representations (for example a program and its flowchart, or a VLSI layout description and its graphic representation) corresponding to a unique abstract one.

- When working *directly* on a common standard representation, the tools may also share the primitives that manipulate it, thus factoring out another component of traditional tool design.

- Tree representations of formal documents (e.g. specifications, programs, proofs) are a long time accepted basis for formal mathematical definitions. Thus this approach is a good step towards the use of the formal techniques, on which advanced software engineering tools will be based in the future.

The abstract representation of a document is a labelled tree. The label of each node corresponds to a syntactic construction of the language (or formalism) used in the document. Figure 13.1 presents a fragment of a Pascal program both in concrete textual form and as an abstract labelled tree.

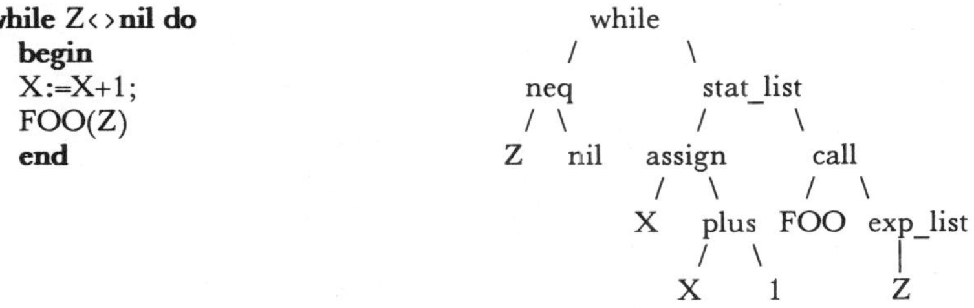

Fig.13.1 Example of abstract representation

Abstract representations are not parse trees. Parse trees often contain structurally irrelevant nodes, and are too dependant on the parsing technique used (because of the constraints it imposes on the defining grammar). The syntactic correctness of an abstract representation of a document in a given language is defined by an **abstract grammar**, which is a tree equivalent of context-free grammars or B.N.F. commonly used to define the concrete syntax of a language. The abstract grammar only reflects the intrinsic structure of the language and is independant of any parsing technique.

An abstract grammar for the language L consists of:

- the definition of the collection of labels (which we call **operators**) allowed in the abstract representation of documents written in L.

- the definition of sets of operators, called **phyla**, corresponding to groups of operators that are allowed in the same context.

- the definition for each operator of a rule specifying its arity (i.e. number of sons) and the phylum associated with each son.

In figure 13.2 we give an example extracted from the abstract grammar of Pascal. The definition of the phylum *STAT* gives the set of operators that may appear on top of a tree to be considered as a statement. The definition of the operator *while* specifies that the sons of a while must be respectively an expression and a statement in that order. List operators may have any number of sons, all belonging to the same phylum.

Fixed arity operators:

intcst	-->	{*integer*}	(* atomic operator*)
ident	-->	{*string*}	(* atomic operator*)
nil	-->		(* atomic operator*)
neq	-->	EXP	EXP
plus	-->	EXP	EXP
assign	-->	VARIABLE	EXP
call	-->	IDENT	EXP_LIST
while	-->	EXP	STAT

List operators:

exp_list	-->	EXP	...
stat_list	-->	STAT	...

Phyla:

VARIABLE	::	ident	meta	unref	index
		dot			
EXP	::	ident	meta	intcst	alfacst
		charcst	nil	hexcst	realcst
		setof	not	uplus	uminus
		unref	hexa	eql	lss
		gtr	neq	leq	geq
		in	intdiv	mod	div
		mult	plus	minus	or
		and	index	dot	format
		call			
STAT	::	meta	goto	repeat	assign
		call	case	while	with
		labstat	stat_list	for	if
EXP_LIST	::	exp_list	meta		

Fig.13.2 Excerpts from the abstract syntax of Pascal

14.2.3. Combining languages and combining documents.

A language called **Metal** [19] may be used to specify new languages to Mentor. The specification of a language L is a Metal program that has to define at least the abstract grammar of L. A Metal specification may contains other components, the most usual being the specification of the concrete textual syntax of the language, and of the translations between concrete and abstract representation. Work is under way to provide in Metal the means to specify more semantical properties of a language, such as scope or type checking [4].

The Metal specification of a language may be compiled into a collection of tables (or virtual machine code) that are used by Mentor to correctly handle this language. Actually Mentor can handle several languages simultaneously, which is a necessary step towards the maintenance of a project data base.

The organization of a data base of project related documents has not yet been much developed in Mentor. However two mechanisms are available that are intended to facilitate the creation of such a data base. They are called respectively **annotations** and **gates,** and they may be used to combine documents or fragments of documents belonging to different languages, or even to create multi-lingual documents.

Annotations are essentially a means of attaching additional information to a node of the abstract representation. This information may just be comments in the usual sense. It may also be any kind of data that may play a role in some circumstances, but is not an essential component of the document: footnotes, references, assertions, examples, etc... In the new Lisp implementation of the system, annotations may also serve as attributes (in the sense of attributed grammars [28]) used to perform computations on the document.

Gates concern the association of values to the atomic nodes (i.e. nodes without sons) of abstract representation trees. Most atomic nodes, in addition to their label, e.g. *ident* or *intcst,* must have an associated value of an appropriate type, e.g. a string or an integer (see fig.13.2). This value may also be an abstract representation tree, belonging to the same or to some other language.

A detailed discussion of gates and annotations is given in [8]. So far they have only been used to produce single documents integrating components in different languages. Annotations or atom values could also be references to fragments of (or places in) other documents, and could thus serve to establish finely grained relations between documents within a project data base.

14.2.4. Manipulation of documents.

14.2.4.1. The manipulation primitives.

The main purposes of the abstract representation of document is to emphasize their internal structure and thus support their manipulation in terms of this structure. Naturally Mentor provides a collection of primitive actions to explore or modify the abstract representation trees. These actions take as argument **cursors** denoting positions in the trees. New cursors may be created as needed, thus permitting to refer simultaneously to several places in one or several documents.

The main classes of manipulation primitives are:

- *local tree navigation* : up, down, right, left to explore in arbitrary order the branches of the tree; next to explore in prefix order; primitives to move onto annotations or gates values.

- *simple tree editing* : modification, deletion, insertion, exchange of subtrees; modification, deletion or addition of new annotations.

- *pattern matching based actions* : creation of incomplete trees to be used as patterns, pattern matching and instantiation, pattern directed tree searches, pattern directed tree traversal and modification. Tree pattern matching is a powerful technique in all symbolic manipulations, from simple interactive structured editing to much larger applications such as pretty-printer generators or source to source program tranformers.

- *parsing and unparsing primitives* : they perform translations between the concrete textual representation and the abstract tree representation; they are driven by tables generated by the Metal compiler from the specification of the language used.

- *filing primitives* : they store and retrieve abstract representation of documents in the host file system.

- *miscellaneous primitives* : for example to get the label of a node, or the number of sons of a list node.

With respect to filing, let us recall that the abstract representation has been defined here abstractedly, i.e. without any thought about its actual implementation. Since the abstract representation is to be the standard one, its is essential that documents be kept in this abstract form at all times, whether in main memory or on file. Furthermore the abstract representation often resolves ambiguities present in the concrete one (such as the association of a comment with the commented text fragment). Thus two equivalent implementations of the abstract representation are actually provided: one in main memory implemented with pointers for fast manipulations, and one linearized (in polish form) in ascii files. Filing primitives perform the translation between the two forms. For a Pascal program, the linearized abstract representation file is about 40 percent shorter than the equivalent text file.

14.2.4.2. Mentor as a syntax directed editor.

Mentor may be used as a syntax directed editor either on a teletype or in full screen mode on a CRT. Teletype mode is convenient when communicating with the system on a low bandwidth line. A bit-map and mouse version is under development.

In CRT full screen mode, the screen is divided into four windows. The largest one contains the visible (pretty-printed) fragment of document currently edited. This fragment corresponds to a tree denoted by a special system cursor: the visualization cursor. The subtree on which commands are actually activated is displayed in bold-face (or underlined characters) within the visible context. It is denoted by an activation cursor. A small window indicates the language and the operator label of the activation subtree. The other two windows are for system messages and diagnostics, and for user input of commands. A fifth window containing help messages may appear when requested by the user. To accomodate window size, it is possible to display abbreviated version of documents, or to scroll the documents in the display window.

New documents may be entered in Mentor either via a multi-entry parser (from a file or from the keyboard) or incrementally with the assistance of a menu system [24]. They may of course be loaded directly in tree form from abstract representation files.

Editing commands are entered with an interactive tree editing language called **Mentol** that can activate the manipulation primitives described above.

14.2.4.3. Creation of new tools.

The Mentol language, though only interpreted, is a full programming language specialized for tree manipulation applications. Its programmability can be used to extend the collection of editing functions, or to develop more complex applications as presented in the next section.

The manipulation primitives are actually Pascal procedures. Thus document manipulation or analysis tools may also be written in Pascal using these primitives, and then linked with the system. Data and control may be exchanged between compiled Pascal code and interpreted Mentol code, allowing one to take advantage of the speed of the former and of the flexibility and interactivity of the latter. As a result, applications written in Pascal may be called interactively during editing sessions.

14.2.5. Practical applications of Mentor.

The primary use or the Mentor system is as an interactive stuctured editor. A variety of languages have already been specified in Metal for this only purpose (e.g. Pascal, Ada, Chill, Metal, Typol, Estelle [16]). The collection of manipulation primitives being limited to purely syntactic manipulation, it is convenient to develop for each language a collection of small or large Mentol programs incorporating some knowledge of the semantics of the language, to assist the user more effectively. Such specialized interactive environments have been developed for Pascal [21], Ada [9], LTR-2 and LTR-3 [34], CPL1.

A special mention is deserved by the language Rapport [22], which has been designed together with its environnement to assist the production of technical reports.

The tools included in these environments may cover many topics:

- *simple navigation and modifications* : for example to find the declaration of an object with a given name, or to automatically create this declaration without actually moving the editing cursor to the declaration part.

- *debugging* : they are based on mechanical insertion (and removal) of debugging statements in the source code.

- *program analysis* : e.g. to produce cross-reference tables, to discover unused segments of programs, or to detect aliasing and potential side-effects [26].

- *program documentation* : automatic adjunction of comments on the basis of analysis results.

- *checking* : tools that ensure the respect of standard semantics (e.g. scope and type discipline), or of local programming norms such as restricted use of the goto statement.

- *static and dynamic measures* : static metrics are easily computed with a structured form or programs [30] Dynamic measures are obtained by mechanical insertion of measuring statements in the source code.

- *automatic program generation* : by means of a library of program skeletons to be filled in with computer assistance.

A larger application of Mentor is the mechanisation of the tranport of Pascal programs between systems and dialects [7,12]. This is a non trivial task, mainly because of the widely diverging conventions concerning separate compilation.

Mechanical generation of these tools for particular languages from a specification expressed in an extension of Metal is an important topic for current research [10].

Mentor is also often used as a tool to assist language design. A extension of Pascal with a good module and interface facility has been designed under Mentor, and a compiler from the extended language to standard Pascal source code was implemented in Mentol [23].

14.3. IMPLEMENTATIONS ISSUES

The choice of Pascal as the implementation language was in many respects very beneficial. Despite some odd and irksome limitations (on the type of function results for example) the language is reasonably well structured and quite readable. It has become at the same time very popular and standardized enough to allow the port of Mentor on a variety of machines and systems without impairing efficiency. As a well structured language, Pascal was a good experimental subject for the techniques developed in Mentor, and the use of the system for its own development has been an unvaluable source of improvements and new ideas, as well as a benchmark for validating those ideas.

However the development of Mentor in Pascal became increasingly difficult as the system evolved. This was due in part to the natural aging of code in an evolving system, but several problems are traceable to the inadequacy of Pascal for the programming structures required by the technology being developed.

14.3.1. Memory management.

Abstract syntax trees are implemented in main memory by means of records linked by pointers. This implementation is storage costly, but it is very dynamic and allows the fast response required by interactive systems. The records have to be allocated dynamically as trees are created but , as is commonly the case is symbolic manipulation systems, there is no way of knowing when a tree ceases to be usable (i.e. accessible) from within the system. Without some form of garbage collection, accessible free storage is slowly used up, and sessions have to be abandonned after a while for lack of available memory. Practically none of the existing Pascal compilers offers a garbage collector (which it is possible although difficult to implement).

A second problem is related to the possibility of dynamically allocating arrays with a size computed at run time. The code of Mentor is independant of the languages edited and all component processors are driven by language specific tables which are loaded when needed. Efficiency of access usually requires the use of arrays to store these tables. The size of each table varies widely according to the complexity of the concerned language. Since the typing structure of Pascal prevents allocation of arrays whose size is computed at run time, the choice is either to choose a maximum allowed size which is usually wasteful and limits the complexity of processable languages, or to handle storage management explicitly at the (considerable) expense of readability and maintainability.

14.3.2. Type discipline.

The very strict and limited type discipline of Pascal has been a hindrance or a source of inefficiency in other ways. One example is that the system needs identical hash-coded tables of objects of different types, but same physical size (pointers): Pascal requires duplication of the source code to avoid typing conflicts.

More importantly, some structures of the system should be type independant if the system is to be open and extensible [14]. This is for example the case with the annotation mechanism. In Mentor, annotations are always trees belonging to some language. Since annotations are very similar in organization to attributes [28], several

Mentor users have requested the possibility of extending them at will to other types and actually use them as attributes. Unfortunately, as with hash-coded tables, their is no simple way to have a unique set of primitives to handle annotations of any type. To some extent the use of variant records, with the discriminant field playing the role of a type tag, may help simulate dynamic typing (i.e. run-time use of values explicitely tagged with their type). Nevertheless, the introduction of any new type requires modification in the source code of the declaration of the variant record type, and the addition of a new case in all subprogram accessing values of that variant record type.

Some of these typing problems could be somewhat alleviated with a more powerful type discipline as in ML [25], but true extensibility of the system by simple addition of new modules without modifications of existing code, requires some form of dynamic typing à la Smalltalk. Of course, a weak alternative is no type discipline at all.

14.3.3. Functional parameters and variables.

Functional or procedural parameters are essential in symbolic applications. In Mentor, for example, one frequently uses tree-walking functions that (roughly) take as argument a tree and a procedure, and call this procedure at every node of the tree. The use of pattern-matching also often requires passing to the matching function an environment-modifying function that can modify the values of variables in the appropriate environment according to the structure of the pattern. A third example is passing the error processing routine as parameter to the system processors in order to have a uniform, though may be user dependant, treatment of error messages.

Although theoretically available in Pascal, functional arguments are usually unreliable or even unavailable in many Pascal compilers, and they were not used in the implementation of Mentor. The tree-walking functions used in Mentor were all written in Mentol.

With respect to the previous section, let us also note that one can simulate dynamic typing in a language having storable typeless pointers and functional values. This is achieved by tagging values with a pointer to a type descriptor containing in its fields the functions (i.e. methods in Smalltalk) associated to the type.

14.3.4. Modularity and separate compilation.

Organization, maintenance or reconfiguration of a large system written in Pascal is rather difficult for lack of a module facility. This is somewhat compensated on most Pascal systems by some form of separate compilation, which may be used to physically separate the logical components of the system. However separate compilation is non standard in Pascal, and it has been diversely implemented. Its necessary use has been a major source of difficulties when transporting Mentor onto new Pascal compiler.

14.3.5. Naming.

Another major maintenance problem was the lack of a good naming facility. Many Pascal implementations do not allow more than 8 to 10 significant characters for identifiers, making it rather difficult to have meaningful but different names for related entities (e.g. the print functions of all types, or the functions giving the language of a tree, an operator, a phylum, etc.). Futhermore, one often wishes to use the same name for entities playing related roles in the program. Standard solutions exist, such as prefixing the name of an object with the name of the module where it is defined (e.g. Ada [33], or overloading a name with several meanings and then determining from the context the intended meaning of each use of this name.

This context may be static as in Ada, or dynamic as in Smalltalk.

14.3.6. Exceptions.

The lack of an exception facility in Pascal has been felt on two accounts:

- exceptions are essential for the robustness (i.e. resistance to failures of all kinds) of a large system,

- exceptions are convenient for the backtrack programming style which is often used in symbolic computation.

Exceptions are available in Ada and Pl/1 in a form oriented towards fail-safe programming, and in Lisp and ML more oriented towards backtrack programming.

14.3.7. Dynamic linking.

Modern software development environments (e.g. Smalltalk, Emacs, Interlisp [11,31,32]) take for granted the possibility of tailoring and extending the environment from inside, that is by adding new code module while the system is running. This dynamic linking facility is usually provided by a language specific environment (it is standard in interpreted languages), but it may also be a characteristic of the operating system (e.g. Multics [27]) and it is not incompatible with compilation (e.g. Le_Lisp [3]).

Dynamic linking is essential to users who want to program specific manipulation routines for immediate use within their session. Its absence in almost all Pascal implementations, together with some of the afore mentionned limitations of Pascal, is one reason for developing Mentol into a programming language. Extensions to Mentor may be programmed either in Mentol or in Pascal, but only Mentol allows dynamic changes.

14.4. THE LISP IMPLEMENTATION

14.4.1. Motivations.

The increasing complexity of the implementation of Mentor, and the limitations of Pascal led us in 1983 to consider starting a new implementation in a different language. Several languages were considered, the main contenders being ML [25], Ada [33], C [20] and some versions of Lisp. The selection was reduced to C and Le_Lisp [3] for several reasons:

- they were readily available and efficiently implemented on many systems,

- they have a very lax type discipline,

- they allow functions and procedures as storable values.

The motivation in the last two points was to have the freedom to choose the most appropriate programming style. The price to be paid is the absence of type checking assistance during program development, particularly in the case of Lisp since limited type checking is available for C [18].

The final deciding factors in favor of Le_Lisp were its memory management with garbage collection, its exception handling primitives, and the existence on top of Le_Lisp of a portable extension for object oriented programming called Ceyx [15].

Augmented with Ceyx, Le_Lisp is a powerful system and our technical choices were validated by the fast realization, the structural simplicity and the conciseness of the new implementation. There were however some drawbacks:

- lack of system assistance with respect to syntax and type checking,

- insecurity and lack of modularity due to the dynamic binding discipline of classical Lisp,

- cost of function calls also due to dynamic binding*,

- higher storage costs (than in Pascal or C) due the use of binary lists rather than records (Le_Lisp does have records, but the space overhead required by garbage collection and compaction limits the usefulness of small records).

- communications between Lisp and other languages are often difficult or limited, especially the call of Lisp functions from other languages.

These remarks are of course specific of Le_Lisp and the situation may be different with other dialects. Le_Lisp itself is being modified to converge towards the Common_Lisp standard which uses a static binding discipline.

Our decision to implement Mentor anew coincided with the start of the pilot project Concerto whose objective is to produce an IPSE prototype for the French PTT. They had arrived independantly to the same choice of implementation language and decided to use this new implementation as the syntactic component of their IPSE.

14.4.2. Organization.

The experience acquired with the Pascal implementation of Mentor has established that its role as a suplier of basic building blocks for the construction of larger tools is as important as its interactive use as an editor. Consequently, it was decided to separate the design of the kernel system from that of the man machine interface. Actually the author has been working exclusively on the design and implementation of this kernel. A man machine interface, including a pretty-printer generator, is being developed independantly within the Concerto project on the basis of the specification of the kernel.

The kernel Lisp implementation is essentially a collection of type, or classes in the Smalltalk sense. The style of programming is essentially object oriented, although types tags are kept at run-time only when necessary so as to reduce space requirements. For each class one must define the representation of values in that class in main memory and in secondary memory, and also define all the functions acting on values of the class. No use is made of the subclassing and inheritance mechanisms available in Ceyx, which was used mainly for its convenient notations and for its record definition facility.

This new implementation extends in several respects the functionalities of the Pascal versions, and especially the pattern matching facilities and the use of gates and annotations. It has a parser based on a Lisp version of Yacc [17] and a Metal compiler. However its abstract representation files are fully compatible with those of the Pascal version when the extensions are not used.

The size of the code is currently on the order of 6500 lines of code (comments included), i.e. less than a third of the corresponding part of the Pascal implementation, although a precise comparison is difficult.

The kernel has been used for a year and a half by about fifteen implementors within the Concerto project for the construction of a complete IPSE. It is a basic

---

* Strangely enough, classical Lisp with dynamic binding is an inefficient language for functional programming.

component of the Esprit project GIPE (Generation of Interactive Programming Environments).

14.5. CONCLUSION

It is our thesis that a software project is essentially the manipulation of a large number of structured documents. The kernel of any IPSE must contain a powerful symbolic computation facility for efficient processing of these documents in terms of their structure. Such a symbolic computation facility has proved, in our experience, to be a sound common basis on which to build a variety of tools that are to cooperate, and it should provide a uniform methodology covering the whole software life-cycle. Futhermore we expect this approach to benefit from the considerable on going research in symbolic computation.

On the pragmatic side, symbolic manipulation cannot be efficiently implemented in all programming languages. Some basic requirements such as efficient memory management or higher order functions disqualify many traditional languages such as Fortran, Pascal or Ada. This is not to say that symbolic computation is impossible with these languages, it is just a lot more difficult. Futhermore the emergence of efficient portable implementations of suitable languages such as Lisp or ML makes them very good candidates for the realization of industrial IPSE's.

Acknowledgements: The development of Mentor and of the ideas it embodies was conducted by the Croap project at INRIA. The first specification of the Lisp implementation was clarified through several discussions with A. Conchon, V. Donzeau-Gouge, G. Kahn, B. Mélèse and Y. Rouzaud. The members of the Concerto project have provided valuable feedback as first users of this new implementation, which is the joint work of the author and P. Borras.

References.

1. Ambriola, V., Kaiser, G.E., and Ellison, R.J., august 1984, *An Action Routine Model for Aloe*, Report CMU-CS-84-156, Carnegie-Mellon University.

2. André, E., Moreau, B., and Rougeot B., 1984, *Vers un atelier flexible et intégré de logiciel: le projet Concerto*, L'Echo des Recherches, 115, 11-20.

3. Chailloux, J., Devin, M., and Hullot, J.M., 1984, *Le_Lisp, a portable and efficient Lisp system*, Proc. 1984 ACM Symp. on Lisp and Functional Programming, Austin, Texas, 113-122.

4. Despeyroux, T., june 1984, *Executable Specification of Static Semantics*, Lecture Notes in Computer Science, Vol. 185.

5. Donzeau-Gouge, V., Kahn, G., Huet, G., Lang, B., and Levy, J.J., 1975 *A structure oriented program editor:a first step toward computer assisted programming*, Proc. International Computing Symposium, North_Holland.

6. Donzeau-Gouge, V., Kahn, G., Lang, B., Mélèse, B., and Morcos, E., sept. 1983, *Outline of a tool for document manipulation*, Proc. of IFIP 83 conf., Paris, R.E.A. Mason (ed.), North Holland.

7. Donzeau-Gouge, V., Lang, B., and Mélèse, B., march 1984, *Practical Applications of a Syntax Directed Program Manipulation Environment*, Proc. 7th Conference on Software Engineering, Orlando.

8. Donzeau-Gouge, V., Kahn, G. , Lang, B., and Mélèse, B., april 1984 *Document Structure and Modularity in Mentor*, Proc. ACM SIGSOFT/SIGPLAN Soft. Eng. Symp. on Practical Software Development Environments, Pittsburgh.

9. Donzeau-Gouge, V., Lang, B., and Mélèse, B., may 1985, *A tool for Ada Program Manipulations: Mentor-Ada*, Proc. International Ada Conference, Paris.

10. GIPE, 1984. *Generation of Interactive Programming Environments.* Esprit Project Proposal, N.348.

11. Goldberg, A., and Robson, D., 1983, *Smalltalk-80. The Language and its implementation*, Addison-Wesley.

12. Hanout, J.C., Kaiser, C., Lucas, H., and Martin, B., 1983, *Experiences in Programming and Transporting Pascal Systems,* TSI - Technique et Science Informatique, 2, 401-417.

13. Henderson, P.B., Edit., april 1984, Proc. ACM SIGSOFT/SIGPLAN Software Engineering Symposium on Practical Software Development Environments, Pittsburgh.

14. Hewitt, C., april 1985, *The Challenge of Open Systems*, Byte, 10, 223-242.

15. Hullot, J.M., 1983, *Ceyx, a multiformalism programming environment*, Proc. of IFIP 83 conf., Paris, R.E A. Mason (ed.), North Holland.

16. ISO/TC97/SC21, june 1985, *Estelle - A Formal Description Technique*, ANSI N 42, DP9074.

17. Johnson, S.C., january 1979, *Yacc, Yet Another Compiler-Compiler*, Unix Programmer's Manual, 7th edition, Bell Telephone Laboratories.

18. Johnson, S.C., january 1979, *Lint, a C Program Checker*, Unix Programmer's Manual, 7th edition, Bell Telephone Laboratories.

19. Kahn, G., Lang, B., Mélèse, B., and Morcos, E., 1983, Metal: a formalism to specify formalisms, Science of Computer Programming, 3, North Holland, 151-188.

20. Kernighan, B.W., and Ritchie, D.M., 1978, *The C Programming Language*, Prentiss-Hall, Englewood Cliffs, New Jersey.

21. Mélèse, B., 1981, *L'environnement Pascal*, INRIA Technical Report N.5.

22. Mélèse, B., june 1984, *Structured editing - Unstructured editing, Cooperation and complementarity*, Actes du 2e Colloque de Génie Logiciel, Nice.

23. Migot, V., sept. 1983, *Un Pascal modulaire sous Mentor*, Thesis, Université Paris XI, Orsay.

24. Migot, V., *A new User Interface in Mentor: the Menu Mode*, Rapport INRIA (to appear).

25. Milner, R., august 1984, *A proposal for standard ML*, Proc. 1984 ACM Symp. on Lisp and Functional Programming, Austin, Texas, 113-122.

26. Morcos-Oury, E., 1979, *Etude des effets de bord des appels de procédures et de fonctions dans le langage Pascal*, Thesis, Université Paris XI, Orsay.

27. Organick, E.I., 1972, *The Multics system: an examination of its structure*, MIT Press, Cambridge, Mass.

28. Reps, T., january 1982, *Optimal-time Incremental Semantic Analisys for Syntax Directed Editors*, Proc. 9th Annual Symp. on Principles of Programming Languages.

29. Reps, T., august 1982, *Generating Language Based Environments*, Tech. Report 82-514, Cornell University ,Ithaca, NY.

30. Schroeder, A., 1983, *Integrated program measurement and documentation tools*, INRIA Research Report N. 227.

31. Stallman, R.M., june 1979, *EMACS: The extensible, customizable, self-documenting display editor*, MIT, AI Memo 519, Cambridge, Mass..

32. Teitelman, W., october 1978, *Interlisp Reference Manual*, Xerox PARC.

33. U.S. Department of Defense, january 1983, *Reference Manual for the Ada Programming Language*, ANSI/MIL-STD-1815 A.

34. Verove, D., march 1983, *Mentor-LTR: Système de manipulation de programmes LTR-V3*, Technical Report DRET-SEMA.

Chapter 15

Advanced support environments;
an industry viewpoint

R.A. Snowdon, N.C. Munro, N.W. Davis and M.I. Jackson

15.1   INTRODUCTION

In the United Kingdom, the Alvey Directorate has been established to direct a coordinated programme of pre-competitive, collaborative research and development in key areas of information technology. A Software Engineering Strategy has been published (1) which focuses attention on areas such as formal methods for system development and metrics for quality and reliability. To help achieve the general goals of improved quality and productivity attention is focused on the concept of the Information System Factory (ISF), a facility enabling the effective production of future information systems. The Alvey Software Engineering Strategy (1) envisages the development of three generations of Integrated Project Support Environment (IPSE) as the means of attaining the goal of the ISF. IPSE 1 is, broadly, the environment provided by UNIX today utilising files and sets of tools. The second generation is seen as employing database technology, rather than elementary filestore and should solve the problem of providing distributed access to the support system. The third generation is expected to make considerable use of methods derived from IKBS as well as provide support for the advanced capabilities to be developed from work on formal methods and metrics. However, although (1) lays out a broad strategy it lacks much detail. If the goals of the overall software engineering programme are to be achieved then it is clearly necessary that the various developments foreseen are set in the appropriate context of the information systems <u>industry</u>. As an IPSE is the means by which industry is provided with a cohesive set of advanced facilities then it is appropriate to understand what are the requirements for such IPSE's from an industry viewpoint. Experience tends to show that the take up of new software methods by the industry is relatively slow. There are, of course, many possible reasons for this (including such as capitalisation) but in the large these derive on the one hand, from a lack of understanding of what a company can obtain from utilising new techniques and on the other from a lack of understanding of what problems industry faces. It is not easy to persuade a company to invest in new technology without some evidence that the investment is going to pay considerable (and, often, short term) dividends.

Although useful, it is not constructive simply to understand the variety and diversity of requirements that the information systems industry has for the development of new technology and tools. Such information should be used together with information on how technology is likely to develop in order to determine how those needs may be best satisfied. This paper reports on some of the results of a study carried out by the authors into the issues concerning the development of advanced IPSEs. This study was carried out specifically in the context of the Alvey Software Engineering Programme with particular stress on the viewpoint from industry. The complete report of the study (2) seeks to identify both the needs for advanced support environments and those research and development activities seen as necessary for environments having the appropriate characteristics for future industrial application. The subject matter of the study was large. Rather than take a narrow but deep view it was considered appropriate to maintain some width in what was considered, with the acceptance that this would, of necessity, be relatively shallow.

15.2   BACKGROUND

The Information Systems Factory (ISF) is a "fully integrated project support environment" (1). Thus IPSE and ISF are seen as the means by which the information systems industry (and by implication the software industry) is enabled to produce high quality products in highly productive ways. What are highlighted therefore are quality and productivity in an industrial setting.

The concept of ISF is explained in (1) by comparing the present development of information systems to a non capital intensive cottage industry. Information systems are produced by the industry by a _process_ involving such activities as determination of market requirements, determination of product strategy, design and development of product, marketing and sales and support, and product enhancement, supplemented by normal business practices such as financial control, resource control and project planning. The effectiveness of this process, as seen from a wide variety of viewpoints, is critical to the success of individual business within the industry. Many factors contribute to the effectiveness of the process, some peculiar to the information systems industry (as opposed to, for instance, the car industry) and some less so. An Information Systems Factory, and thus an IPSE, is a manifestation of the process. Through such environments the industry is provided with techniques, tools and ways of working which enable it to be effective and to produce quality products. IPSE is thus the context of the

process. Of course an IPSE may change (improve) the
process, so it is necessary to retain a high level or
abstract view of what is involved in the process as well
as taking detail into account.
     This "definition" of IPSE is extremely broad. It
includes, for example, factors concerned with the physical
attributes of the environment (e.g. office design, desks)
and even social and organisational considerations. This
definition is beyond the scope of the normally accepted
definition of an IPSE as being a computer based set of
facilities supporting the software process. The study,
and this paper, concentrates on this more limited view
within the broader definition, as does the Alvey Software
Engineering Strategy.
     There is, of course, no unique software process.
Different types of products have different requirements.
Likewise, different companies within the industry work in
different ways or place different emphases on activities
within the process. It is clear, therefore, that there is
no such thing as a single IPSE to be used in every
circumstance. There is, of course, much similarity
amongst the variety of processes and of requirements and
hence it is reasonable to consider such concepts as the
Kernel IPSE and generic IPSE capabilities.
     IPSE is a relative concept. There are already, by
definition, in existence environments which enable (to
greater or lesser extents) the production of software
systems. None however could really be described as an
<u>integrated</u> environment supporting all aspects of the
process. Some considerable work has resulted from efforts
associated with the Ada programme to produce an Ada
programming environment. However none of this work has
yet resulted in an advanced environment suitable for the
support even of programming activities. The CADES system
(3) provides a specialist environment supporting a large
group of people working on the development of ICL's
mainframe operating system. More recently Apollo Computer
have constructed the DSEE system (4) based on their range
of networked workstations. Other systems (e.g. SDS (5)
and Perspective (6)) have been produced to support
specific aspects of a software project. It is clear,
however, that none of these systems comes anywhere near to
the concept of the fully integrated IPSE leading to the
ISF.
     Each IPSE is, of course, a computer system.
Improvements in computer systems technology make
available, and raise expectations of, new facilities for
the users of an IPSE. The existence nowadays of powerful
personal workstations having good man-machine interface
characteristics has demonstrated the increased
effectiveness available to programmers and others.
Systems such as the Cedar environment at Xerox PARC (7)
provide pointers to what systems in industry may be able
to rely on in the future. Such systems are, at present,
expensive, so there is a need both to predict future cost
reductions and also to be able to determine the likely

return to be achieved from investment in such systems.

15.3   THE STUDY

Against this background, the authors (forming a team from ICL, STC IDEC and STL), carried out a study under a contract with the Alvey Software Engineering Directorate into the whole area of future integrated project support environments. The study was carried out over a period of about 4 months in the late summer of 1984. The final report from the study is available as reference (2).

The objective of the study was to further define the requirements for, characteristics of, and research and development activities required to enable the development of advanced support environments of relevance and value to the future information systems industry.   The specific framework for advanced environments was that described in the Alvey Software Engineering Study (1) - namely of IPSE 2 and IPSE 3 leading to the Information Systems Factory.

The study team visited 40 companies and institutions in the UK, USA and mainland Europe receiving input from about 96 individuals.   A two day meeting was held towards the end of the study involving the project team, external consultants and members of the Alvey Software Engineering Directorate.   At this meeting the interim results of the study were presented and detailed recommendations for future activities in the development of advanced support environments discussed in detail.   The final report (2) reflects these discussions.

The remaining sections of this paper report on the major issues highlighted by the study.   It is emphasised that the issues discussed are raised primarily from the point of view of perceived industrial needs rather than being of a purely technical nature.   Of course technical issues   are   raised,   but   within   the   framework   of requirements.

A final section of this paper gives a brief summary of the projected characteristics of 2nd and 3rd generation IPSE's.   Reference (2) contains much more detail on this subject.

15.4   ISSUES

The aim of an integrated project support environment is to provide a computer based set of facilities which enables a company to carry out its business, of producing information systems, in highly effective ways.   The systems produced should be competitive both in terms of quality and in terms of cost and availability.   An information systems company will need to invest in the technology represented by future IPSEs if it is to remain competitive.   The basic issue to be faced, therefore, is concerned with how much investment for what return.   This question cannot be answered directly for a number of good reasons (e.g.   the difficulty of predicting the value of new techniques, the variety of needs within the industry).

It is possible, however, to consider a number of issues at
a slightly lower level of abstraction, each of which has a
bearing on the issue of cost effectiveness. This is the
viewpoint taken in the following sections. In each of the
subsections belows, technical issues are raised in
response to a more general issue.

15.4.1 Gearing

The facilities provided by an IPSE should act to
improve the effectiveness of a company in producing
information systems. The IPSE should therefore provide a
gearing effect.

Three levels of gearing are identified.

(i)   Individual

(ii)   Project

(iii)   Corporate

In order to discuss these in a little more detail a simple
model for describing the development process is
introduced. This is shown graphically in figure 15.1.

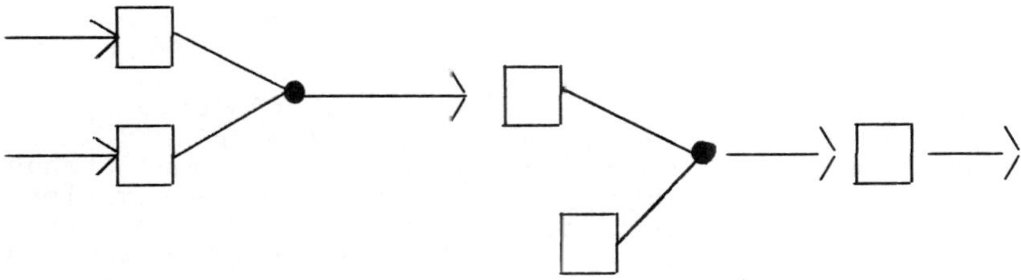

Figure 15.1

In this model, the development process is seen as being
represented by a set of activities (represented by ●)
operating on objects to produce objects (represented by
□). The objects presented in the process will often be
documents, but more abstractly can be thought of as
elements of a design or of an information system. The
activities may be relatively formal (e.g. "compile a
program" or "carry out a test") or relatively informal
(e.g. "obtain user requirements", "conduct market
survey"). Of course, this model is overly simple, but it
suffices for the present purpose.
        Individual activities within the model will be
carried out by some agent. In many cases this agent will
be a person. The individual gearing that can be provided
by an IPSE is concerned with enabling a person carrying
out activities to be more effective. Measures of
effectiveness must include both time and accuracy.

Examples of what can be provided by an IPSE in terms of individual gearing include:-

for a   - a <u>better</u> compiler (faster, better
programmer   diagnostics etc)
                    - new languages allowing the programmer
                      to express his ideas accurately and
                      easily.
                    - new methods, allowing the programmer to
                      produce better implementations

for a   - a database system to provide effective
personnel   access   to   personnel   data
manager   - expert systems to allow skills matching
                      to be carried out

for a   - good   communications   facilities
salesman   - databases

for a   - effective library facilities, supported
researcher   by searching, indexing and document
                      retrieval capabilities
                    - computer systems allowing the rapid
                      establishment and evaluation of
                      experimental systems

for a   - control tools
project   - PERT
manager   - spreadsheet

In each case, a person is made more effective by the provision of good tools designed to support a role.

Of course, the provision of good tools may change the activity (or more often a related set of activities) so that it is important to realise that future IPSEs should not simply provide gearing for activities which are currently part of the process. There is a need for future IPSEs to simplify the activities carried out by an individual and hence provide gearing in that way as well as by enabling a person to carry out existing activities more effectively.

A person carrying out a set of related activities is hindered if there are unnecessary obstacles in his way. In terms of an IPSE this has implications, for example, on the coherence of a set of tools (i.e. the way a set of tools fits together) or on the uniformity of representations of data. An illustration of the latter (as an example of unnecessary obstacles) is the variety of formats expected from different document processing systems.

Individual effectiveness, in terms of the use of a set of computer based tools, is significantly affected by the ease of use of those tools. Future IPSEs will, in general, provide an interactive style of working whereby the human user is working closely with the tools provided by the IPSE in a workstation-like environment. It is critical, therefore, that future IPSEs pay appropriate attention to the man-machine interface. This has a number of implications (e.g., size of screens, touch screen, light pen, mouse, use of graphics), not the least of which is to ensure that the individual has sufficient processor cycles available to him with fast access to store. The Xerox PARC mentality that the human user of the IPSE should be "think bound not compute bound" is a significant message for future IPSEs. Of course, underlying such considerations is the basic one of capital cost. Very fast workstations offering powerful man-machine interactive capabilities are still relatively expensive. An issue to be faced, therefore, is the trade-off between the gains to be made by an individual from his use of such equipment and its cost. It may well be the case that different roles will be best served by different types of workstation.

<u>Project gearing</u> is the second level to be considered. A project may be considered as a coordinated set of activities involving a number of individuals, each contributing to the overall aims of the project. Whereas IPSE support for individual roles concentrates on gearing the activities relating to those roles, IPSE support for a project is concerned with providing means by which the set of activities making up that project are made more effective as a whole. The sort of support that an IPSE should give at this level includes .

(i)   Coordination of activities
(ii) Sharing and control of information
(iii) Scheduling of resources

Integration and cohesion of tools and information representation is important in providing smooth ways by which information can flow amongst individuals within a project. Support for project standards (hopefully in constructive ways rather than by negative policing) is needed. A key issue at the project level is that of version and configuration control. One of the largest problems faced by project teams working on software development is to maintain control over the knowledge of which bits fit together and in which versions.

An important "role" at project level is that of the project manager. He needs to be able to keep control over the ongoing state of his project. Thus, the IPSE must provide him with visibility on the state of his project and simplify his job by means of such things as automatic monitoring facilities.

Projects are (generally) comprised of a number of people working together. Future IPSEs must therefore provide the means for effective community working. This has implications in terms of database design (sharing, locking, privacy) which are exacerbated by the use of powerful workstations communicating over networks. Community working also means that IPSEs must be able to offer such things as electronic mail in a properly integrated way.

Whilst future support environments must address issues relating to both individual and project gearing both of these emphases are secondary to the major concern of making a <u>business</u> more effective. Individual gearing affects individual roles. Project gearing provides the means by which individual projects are carried out more effectively. However, projects are not (in general) carried out in isolation. A particular company is effective because of, amongst other things, its ability to capitalise on its experience and specific skills. Thus, there is a third level of gearing which is of extreme importance for future IPSEs, and in particular for the fully integrated ISF. This is the gearing effect at the <u>corporate</u> level.

In a business there will be a number of separate but related projects. The effectiveness of a company in carrying out these projects will depend, for example, upon its ability to control the inter-relationships and cross dependencies amongst them. Such relationships might be technical (e.g. one project relies on the results of another), to do with resourcing (e.g. one project may on rely another project finishing to release key skills) or concerned with business achievements and finance.

In carrying out a project (or even an individual activity) within the corporate context, effectiveness can be affected significantly if the advantages of working in this context are made available. A company adopts standard working practices, methods and tools which are tailored to the business needs of that company in order that it functions effectively. System development activities should therefore be able to take advantage of the larger corporate context through an advanced support environment which recognises this position. A company is able to produce certain products because it has the experience and knowledge of developing these products. The advanced IPSE must respect this by providing means for the corporate "experience-base" to be built up as projects progress and to make available this expertise to subsequent activities. One good example of this is the concept of component reuse. One of the assets of an informations systems company is the products it has already produced. The availability of a corporate component library, together with means of developing subsequent systems using this library, should represent a powerful element in future IPSEs.

Examples of facilities which relate to the gearing at the corporate level include:
- control of inter-project relationships
- construction and utilisation of a corporate "experience base"
- dissemination of standards
- automatic data collection from projects; data analysis

Other issues which relate to the discussion of IPSE at this level include that of the relationship of an IPSE to the variety of capabilities that will exist to help in

the running of any business. These include reporting
systems, payroll and personnel systems, order databases
etc. Further discussion on these points appears under the
"scope" heading below (15.4.2).

<u>15.4.2</u>   Scope

The notion of an integrated project support
environment supporting the effective production of quality
information systems owes something to the concept of an
integrated <u>programming</u> support environment. The scope of
a programming environment is relatively well bounded being
concerned with supporting the various inter-related
activities concerned with program development (e.g.
preparation of program texts, compilation, testing,
debugging). Such environments may support several
programmers or just an individual. The scope of an
integrated <u>project</u> support environment (or an information
systems factory) is significantly greater. The issue that
is raised therefore is what are the bounds of an IPSE, and
more particularly, what are the bounds of IPSEs to be
produced in the relatively near future if the ultimate
scope is that of the whole context of the information
systems development process within an information systems
factory.

Related to this general problem are issues deriving
from the variety of requirements for support environments
within the information systems industry. It is certainly
true that different types of products have different
technological requirements (cf. computer games and
life-critical control systems). It is also time that
different types of companies have different requirements
arising from their different attitudes to business (cf. a
small commercial software house and a large mainframe
manufacturer).

These issues raise questions concerning the
generality and the tailorability of future IPSEs. It is
certainly true to say that it is very unlikely that there
will be one, universally applicable, complete IPSE
providing full integration of all corporate, project and
individual facilities and providing all of the latest
technology. What is needed, therefore, is the development
of an architecture for IPSEs which must provide the
appropriate infrastructure capabilities to enable gearing
at the levels discussed in section 15.4.1. This
infrastructure should enable the construction of IPSEs to
satisfy specific needs by such means as the support of
specialist tools or the tailoring of general purpose
capabilities to suit individual requirements.

An example of scoping can be inferred from the point
of view of a programming environment. A programming
environment supports those roles concerned with
programming. Other environments (e.g. analyst's
environment, specificaton environment, maintenance
environment) concerned with other roles can be envisaged.
An IPSE could then be constructed by combining these

various smaller environments. How well such a conglomerate might function (i.e. the degree of integration) would depend on a variety of factors of course.

Another, orthogonal view, might be called "scoping by generic process'. The various roles and effects to be supported by an IPSE have much in common. What is unique to specific circumstances is the way in which general capabilities are used. Thus, it is possible to consider the construction of IPSEs as being more and more specialisations of general capabilities. Examples include:

- database systems; schema capabilities provide specialisation
- compiler-compilers; providing a means of developing translators for a variety of languages
- structure editor generators; providing the capability for tailoring editors for specific structures
- office systems technology; providing a base for many of the activities carried out by various roles in the development process.

The issues raised for future IPSE development are thus concerned with determining which are the useful limited scopes for the less than complete IPSEs, how the generality can be best provided and how the complete, corporate IPSE can be composed from these smaller parts.

15.4.3   Investment

The introduction of advanced technology, in the form of advanced support environments, implies a need to consider the value or worth of the accompanying investment. The issue raised concerns the fundamental equation of hardware, software and skill costs set against the benefits to a business arising from the adoption of the new technology. Investment considerations need to be set in the context of the current spend by companies on computing capabilities for its staff and the write-off for that investment.

There is a marked shortage of published material containing worthwhile cost-effectiveness figures (But see Dolotta et al. (8), Boehm (9)). It is clear therefore that there is a need for appropriate metrication capability to be put in place as part of an IPSE from the beginning. The Alvey Software Engineering Strategy (1) places strong emphasis on the development of metrics and associated interpretation and analysis facilities.

There is no doubt that there is concern at government level about the need for significant investment in the development of advanced software engineering capabilities for industry. Large sums of money have been provided in the U.K. under the Alvey Programme and in Europe under ESPRIT. In the end, however, it is companies in the information systems industry who will have to make the investment to put the new techniques in place. The present lack of quantifiable measures against which such investment can be judged makes it hard for a company to determine a proper decision. Thus, there is evidence at present that the software engineer is very much the poor relation in terms of investment in comparison to, for example, the VLSI designer. There are, of course, many factors which must be taken into account before a company might invest in an IPSE. These include decisions on the scope and the type of equipment to be used, decisions relating to the corporate style and need (e.g. graphical capability might be inappropriate) and on other issues such as legal or security constraints (e.g. paper copies required), the skills and backgrounds of its employees or the degree of geographic distribution of the organisation. The company has to gauge its product costs and the size of its market. Product cost will, in part, depend on IPSE cost, which in turn will depend on the size of the IPSE market, the overhead of providing a tailored IPSE etc.

15.4.4   Adaptability

It is clear that adaptability should be a major concern in the development of advanced support environments. Comment has been made earlier of the fact that the universal IPSE cannot usefully be constructed, the implication being that IPSEs will need to be constructed or tailored according to specific needs. There is an issue here in that, certainly in the short to medium term, it may well be appropriate to consider the development of IPSEs which are aimed specifically at certain kinds of application area. Thus, the different needs of widely disparate parts of the information systems industry might be best served by a (small) number of (relatively) specialist systems. The question raised then is, therefore, "which areas?".

There is another reason why adaptability is important. This is concerned with the evolution and the need to protect investment. Advanced support environments will have to operate in a world where there is considerable existing investment in a variety of incompatible software and hardware systems, early IPSEs, programming and management support environments, office technology and people skills. It is obvious that the new technology represented by an advanced IPSE must relate positively to this investment. Thus, it is a concern for advanced environments that they represent a (relatively) smooth continuum. A programme developing such environments must, therefore, provide an evolutionary capability, each "upgrade" adding to the gearing effects already provided. One particular problem concerns the existing skills base of a company. Advanced environments may well provide greater capabilities for their users by providing new ways of doing things. There is then the problem of ensuring that the user of the environment is sufficiently able to make use of the more advanced capabilities now available to him. One view is that the

IPSE should contain the necessary educational and training capability to help overcome such difficulties.

15.4.5 Availability

In this final subsection is drawn together a set of technical issues concerning the availability of technology for the implementation of IPSEs.

The present day project support environment in industry is generally based on mainframe or mini-computers. There are just appearing workstation networks having sufficient capabilities to offer an alternative to this. The Apollo DSEE system (4) is perhaps the best known of these latter, but other environments based on UNIX, for example, run on machines such as the ICL PERQ or SUN workstations. Systems such as Cedar (3) or Smalltalk (10) also run at advanced laboratories like XEROX PARC. The Ada community are working to provide programming environments for the Ada language, but none is readily available yet. It is generally true to say that existing IPSEs are also characterised by:-
-       non-integrated support for a few roles
-       the use of either filestore or commercial,
        non-distributed databases
-       a lack of machine cycles per person
-       poor man-machine interface
-       file transfer communication over LANs and WANs
-       non-integrated mail
Coming very shortly (and even here in some cases), are:-
-       cheaper workstations with better MMI (but still not
        good enough)
-       (some) more cycles per person per £
-       large components (already here in the commercial
        applications world in some aspects)
-       formally based development methods providing
        opportunities for verification and correctness
-       expert systems, in particular management decision
        support systems.
In the longer term, there are a number of "existing" developments which raise issues concerning the implementation possibilities for future advanced environments. These include
-       Fifth generation computers and improved system
        architectures
-       Developments in IKBS
-       Distributed systems
-       Much better man machine interfaces
-       Significant improvements in communications
        capability

15.5   SUMMARY OF IPSE 2 and IPSE 3 CHARACTERISTICS

The final report of the study (2) lays out in some detail the projected characteristics of second and third generation (IPSE 2 and IPSE 3) advanced support environments. Below is a very short summary taken from (2)

IPSE 2	IPSE 3
- is a pragmatic environment The techniques it provides are not revolutionary. It can be provided on machines and software systems which will certainly be available.	- is an ISF providing significant improvements in productivity and quality by powerful techniques requiring development in order to achieve objectives. These include formal methods and component reuse. May well make significant use of novel architectures and languages.
- is oriented towards the production of software systems	- supports the production of complete information systems
- is CAD for software	- is ISF
- pragmatic approach to <u>software</u> component reuse	- full support for component reuse, including generic capabilities
- is principally aimed at supporting the individual <u>project</u> (and thus project roles) within the industry	- provides support for, and relates to, the <u>corporate</u> requirement of the industry, and thus corporate roles
- is a passive environment in which the user must take the initiative in carrying out any activity	- in an active environment in which actions are carried out automatically without unnecessary action by the user
- the user interface is at the level of a database	- the user interface is defined at abstract, problem-oriented levels
- configuration/capabilities are relatively static, fixed by the IPSE supplier with relatively little tailoring of capability for user	- the configuration and capabilities are dynamic; the user is able to configure his environment to his role and his requirements; IPSE 3 is characterised by its generic capabilities

REFERENCES

1.   Talbot D., and Witty R.W., 1983,
        Alvey Programme Software Engineering Strategy

2.   Advanced Support Environment Study 1985
        UK Department of Industry, to be published by IEEE
        on behalf of the Alvey Directorate

3.   Snowdon R.A., 1981,
        CADES and software system development, in Software
        Engineering Environments, H. Hunke, editor, North
        Holland

4.   McLean G., 1985
        DSEE - Overview and configuration management,
        Chapter 2 of this volume

5.   Software Sciences Ltd., 1982
        Introduction to SDS, Software Sciences Ltd.,
        Macclesfield, UK

6.   Systems Designers Ltd.,
        Perspective, the support environment marketed by
        Systems Designers Ltd., Fleet, Surrey UK.

7.   J. Donahue 1985
        Cedar : An Environment for "Experimental
        Programming",
        Chapter 1 of this volume

8.   Dolotta T.A., Haight R.C., and Mashey J.R., 1978
        The Programmer's Workbench,
        B.S.T.J, July - August 1978

9.   Boehm B.W., 1981
        Software Engineering Economics,
        Prentice Hall Inc., Englewood Cliffs, N.J.

10.   L.P. Deutsch, 1985
        Project Support in the Smalltalk 80 Integrated
        Environment, Chapter 10 of this volume.

Index